PROBLEMS IN KINETICS, PHOTOCHEMISTRY AND KINETICS OF GASES

PROBLEMS IN KINETICS, PHOTOCHEMISTRY AND KINETICS OF GASES

M. G. ARORA

ANMOL PUBLICATIONS PVT. LTD.
NEW DELHI-110 002 (INDIA)

ANMOL PUBLICATIONS PVT. LTD.
4374/4B, Ansari Road, Daryaganj
New Delhi-110 002

Problems in Kinetics, Photochemistry and Kinetics of Gases

First Edition 1997

ISBN 81-7488-564-1

PRINTED IN INDIA

Published by J. L. Kumar for Anmol Publications Pvt. Ltd., New Delhi-110 002 and Printed at Mehra Offset Press, Delhi.

PREFACE

To solve the problems—especially numerical problems—is a good exercise in pure sciences. This enthuses the process of understanding the theory—thus helps learning.

This present book will be much useful for the students of chemistry and Physics. In the Part I of the book more than one hundred problems pertaining to all aspects of Kinetic theory of Gases are given—their solutions are given in part II of the book. Part III contains problems relating to Chemical Kinetics and Photochemistry, while their solutions are given in the last part IV of the book.

We believe this book will greatly serve the purpose of undergraduate and postgraduate students in the concerned fields. Teachers will also find the book most informative.

M. G. Arora

Part I
PROBLEM STATEMENTS

GASES AND KINETIC THEORY

PROBLEM 1. The temperature of a given gas is −10°C. What are the equivalent Fahrenheit and absolute Kelvin scales?

PROBLEM 2. If the temperature of the air in a chemistry laboratory is ambient (77°F), what is the equivalent scale in Kelvin?

PROBLEM 3. The bar is a unit of pressure defined as 1.00×10^6 dynes/cm^2, or 1.00×10^5 N/m^2. What is the equivalent of the bar in units of mm Hg?

PROBLEM 4. The density of liquid gallium is 6.09 g/cm^3 at 35°C. If this element is employed in a barometer instead of mercury, what is the height of a column of gallium sustained in the barometer at 1 atm pressure?

PROBLEM 5. If the height of a water column is 600 mm, what

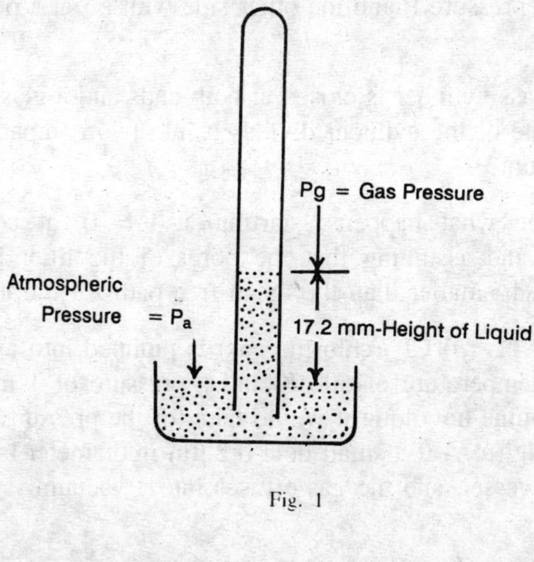

Fig. 1

will be the height of the mercury column?

PROBLEM 6. In a popular laboratory experiment in general chemistry, a gas is collected in an inverted tube. A typical situation at the end of the experiment is shown in Figure 1. If the difference in levels of the liquid surfaces is 172 mm, the specific gravity of the liquid is 1.23, and atmospheric pressure is 763 mm Hg, find the pressure of the gas in mm Hg.

PROBLEM 7. An open and mercurial manometer shown above in Fig. 1 is connected to a gas tank. The mercury meniscus in the left leg of the manometer is opposite 33.8 cm on the meter stick; in the right leg it is opposite 16.2 cm. Atmospheric pressure is 747 mm Hg. Find the pressure of the gas.

PROBLEM 8. If P_a = 755 mm Hg, find the gas pressure in the demonstration setup in Figure 1.

PROBLEM 9. A gaseous mixture of 90 mole percent hydrogen and 10 mole percent deuterium at 25°C and a total pressure of 1 atm effuses through a small pin hole of area $0.30 mm^2$. Calculate the composition of the intial gas that passes through.

PROBLEM 10. One side of a glass cylinder is initially filled with H_2 at a pressure P and the other side with N_2 at a pressure 2P.

This glass cylinder is closed at both ends and a glass filter in the middle of the cylinder divides it into two compartments of equal volume.

Describe what happens, starting at t = 0, at constant temperature and assuming that the pores of the filter have a diameter much smaller than the mean free paths of the gases.

PROBLEM 11. (*a*) Cl_2 (chlorine gas) is pumped into a 1-dm^3 vessel at a temperature of 300°K and a pressure of 1 mm Hg (torr). Determine how long it would take for the pressure to fall to 0.5 mm Hg (torr) if a small hole 0.2 μm in diameter is punctured in the vessel, and the gas effuses into a vacuum.

(*b*) Identify and justify the assumption that you made in the calculation.

PROBLEM 12. Kinetic theory of gases predicts that the viscosity of a gas is proportional to $T^{1/2}\ \sigma^{-2}$ where T is the temperature and σ is the molecular diameter. Give an explanation of why viscosity (*a*) increases with temperature, (*b*) depends on the diameter of the molecules, and (*c*) is independent of pressure at constant high temperatures.

PROBLEM 13. One of the two principal methods for the determination of viscosities of gases is the Poiseuille's flow experiment. In this experiment, a sample of air at 298°K was allowed to flow through a tube with radius 1mm and length 1m. Pressures at both ends of the tube were 765 mm Hg and 760 mm Hg respectively. A volume of 90.2 cm^3 measured at the lower pressure of 760mm Hg passed through the tube after 100 sec. Calculate the viscosity of air at this temperature.

PROBLEM 14. Knudsen effusion technique was used to determine the vapor pressure of scandium and the following data were obtained

temperature, 1555.4°K

time, 110.5 min

weight loss and diameter of orifice were 9.57 mg and 0.2965 cm, respectively.

Use the above data to estimate the vapor pressure of scandium at 1555.4°K and 1 atm.

PROBLEM 15. A sample of beryllium powder was placed in a box having an effusion hole of 3.18 mm in diameter. After 15.2 min and a temperature of 1537°K, 8.88 mg of the powder had effused. Use the method of Knudsen to determine the vapor pressure of Be at 1537°K.

PROBLEM 16. Calculate the average velocity of Caesium atoms forming an atomic beam emerging from a small hole of a wall of an oven. The oven contains a sample of Caesium and is heated to 500° C.

PROBLEM 17. The escape velocity, the speed at which molecules can escape the earth's gravitational field, is approximately 1.1×10^6 cm sec^{-1} (about 7 miles sec^{-1}). At what temperature, approximately, will the root mean square velocity value for H_2 be equal to the escape velocity?

($R = 8.3 \times 10^7$ erg $mole^{-1}$ deg^{-1} ;
at. wt: H =1)

PROBLEM 18. Air is estimated to be 20 mole per cent O_2 and 80 mole per cent N_2 and the total external surface area of an average human being is $2.0 m^2$. Estimate the number of collisions of air molecules per second with the surface of an average human being at 25°C and 1 atm.

PROBLEM 19. A sample of caesium is heated in an oven to 500°C. Calculate

(*a*) the number of collisions a single Cs atom makes inside the oven per second;

(*b*) the number of collisions all the atoms make per second. Assume the volume of the oven to be 50 cm^3, the vapor pressure of Cs at 500°C is 80 mm Hg, and C = 351 m sec^{-1}.

PROBLEM 20. A gas in two dimensions is enclosed in a rectangular container. Calculate the number of molecular collisions on a unit length of the container.

PROBLEM 21. The concentration of hydrogen atoms in intergalactic space at a temperature of 7.8×10^{50} K is about 1 atom per 100 liter. Assuming the collision diameter of a hydrogen atoms is 2.0 Å calculate

(*a*) the mean free path of an atom in light-years.

(*b*) The approximate average time, in years, between collisions of a given atom.

PROBLEM 22. The following equation is given for 1 mole of a gas

$$V = f(P,T)$$

Derive the ideal gas equation.

PROBLEM 23. A sample of an ideal gas at 27°C and 380 torr occupied a volume of 492 cm^3. What was the number of moles of gas in this sample?

$$(R = 0.082 \text{ liter atm mole}^{-1} \text{ deg}^{-1}).$$

PROBLEM 24. Calculate the number of moles of air in a column 1 cm^2 in area and rising from sea level to an altitude of 1.5×10^6 cm (about 10 miles), allowing for the barometric decrease in pressure. Assume air to be a uniform gas of molecular weight 29, and a uniform temperature of 0°C. Note that the atmospheric pressure is equal to the total weight of gas in a column of unit area and of infinite height. Conversion factors: R = 0.082 liter-atm/mole-deg = 8.31×10^7 ergs/mole-deg = 1.98 cal/moledeg; 1 atm = 760 mm Hg = 1×10^6 dynes/cm^2.

PROBLEM 25. A 2-liter Dumas bulb contains *n* moles of nitrogen at 0.5 atm pressure and at T°K. On addition of 0.01 moles of oxygen, it is necessary to cool the bulb to a temperature of 10°C in order to maintain the same pressure. Calculate *n* and T.

PROBLEM 26. Under what conditions will a pure sample of an ideal gas not only exhibit a pressure of 1 atm but also a concentration of 1 mole liter^{-1}?

PROBLEM 27. (a) Show that the pressure of the column of air at the height *x* in Figure 1 is given by

$$P = P_0 \exp\left(-\frac{Mgx}{kT}\right)$$

P_0 is pressure at $x = 0$, g is the acceleration due to gravity and M is the average molecular mass of molecules in air. Assume a uniform air temperature T in the gravitational field of the earth.

(b) What is P when P_0 = 760 mm Hg, T = 273°K and x is 8km?

PROBLEM 28. The stopcock between a 3-liter bulb containing oxygen at 195 torr and a 2-liter bulb containing nitrogen at 530 torr is opened. After equilibration, what is the final pressure? Assume that T is constant.

PROBLEM 29. Two flasks of equal volume are filled with nitrogen gas and, when both are immersed in boiling water, the gas pressure inside the system is 0.5 atm. One of the flasks is then immersed in an ice-water mixture, keeping the other in the boiling water. Calculate the new pressure for the system.

PROBLEM 30. The helium content of the atmosphere at sea level is about 5×10^{-4} mol %. A round 100-cm^3 Pyrex flask with wall thickness of 0.7 mm is evacuated to 10^{-12} atm and sealed; calculate the pressure in the flask at the end of one month due to inward diffusion of helium. At 20°C the permeability of Pyrex glass to helium is given as $4.9 \times 10^{-13} cm^3 s^{-1}$ per millimeter thickness and per millibar (m bar) partial pressure difference.

PROBLEM 31. The gauge pressure in an automobile tire when measured during the winter at 32°F was 30 pounds per square inch (psi). The same tire was used during the summer, and its temperature rose to 122°F. If we assume that the volume of the tire did not change, and no air leaked out between winter and summer, what is the new pressure as measured on the gauge?

PROBLEM 32. A Dumas bulb is filled with chlorine gas at the ambient pressure and is found to contain 7.1 g of chlorine when the temperature is T°K. The bulb is then placed in a second thermostat bath whose temperature is 30°C hotter than the first one. The stopcock on the bulb is opened so that the chlorine-gas pressure returns to the original value. The bulb is now found to contain 6.4 g of chlorine. Calculate the value of the original temperature. If the volume of the bulb was 2.24 liters, what was the ambient pressure? The atomic weight of chlorine is 35.5.

PROBLEM 33. Two separate bulbs contain ideal gases A and B, respectively. The density of gas A is twice of gas B, and the molecular weight of gas A is half that of gas B; the two gases are at the same temperature. Calculate the ratio of the pressure of gas A to that of gas B.

PROBLEM 34 An iron pipe 2m long and closed at one end is lowered vertically into water until the closed end is flush with the water surface (see Fig. 1 below). Calculate the height H of the water level in the pipe.

Miscellaneous data : 25°C, diam. of pipe is 2 in., density of water is 1.00 g/cc, barometric pressure is 1.00 atm. Also, 1 atm = 10^6 dynes/cm^2 = 10 m hydrostatic head of water (neglect the effect of water vapor pressure).

PROBLEM 35. What volume (in liters) will a 0.17 g sample of H_2S gas occupy at 27°C and 380 torr?

PROBLEM 36. Consider the apparatus in the diagram. If the temperature of the oil is raised from 68°C to 136°C, how large is the final volume compared to the initial? (Ignore the small effect of pressure change; *i.e.*, assume that the pressure remains constant.)

PROBLEM 37. A pioneer aeronaut is planning the design of a hot-air balloon. What volume of air at 100°C should be used if the balloon is to have a gross lifting power of 200 kg (defined as the mass of displaced air minus the mass of hot air)? The ambient temperature and pressure are 25°C and 1 atm, and the average molecular weight of air is 29 g/mole, whereas that of the hot air is 32 g/mole (due to the presence of some CO_2).

PROBLEM 38. Air is approximately 80% nitrogen and 20% oxygen (on a mole basis). If 4.0 g of hydrogen is added to a 22.4-liter flask maintained at 0°C and initially filled with air at 1 atm pressure, what will be the molecular weight (*i.e.*, the average molecular weight) of the hydrogen-air mixture?

PROBLEM 39. An 11-liter flask contains 20 g of neon and an unknown weight of hydrogen. The gas density is found to be

0.002 g/cc at 0°C.

Calculate the average molecular weight and the number of grams of hydrogen present, and also the pressure. The atomic weight of neon is 20 g/mole.

PROBLEM 40. Calculate the composition in weight per cent of a mixture of 5.00 g of helium and argon occupying a volume of 10.0 liters at 25°C and 1.00 atm.

PROBLEM 41. Hydrogen gas will dissociate into atoms at a high enough temperature, *i.e.*, $H_2 = 2H$, Assuming ideal gas behavior for H_2 and H, what should be the density of hydrogen at 2000°C if it is 33% dissociated into atoms? The pressure is 1 atm.

PROBLEM 42. Calculate the minimum work that must be expended in the compression of 1 kg of ethylene from $10^{-1}m^3$ to 10^{-2} m^3 at a constant temperature of 300°K, assuming the gas to be

(a) ideal (b) Van der Waals.

PROBLEM 43. A certain substance obeys the van der Waals equation, and its constants *a* abd *b* are known. Name six types of properties or coefficients that may be calculated for this substance, using the above information.

PROBLEM 44. Determine the equation of state for a certain hypothetical gas whose thermal coefficient expansion,

$$\alpha = \frac{1}{V}\left(\frac{\partial V}{\partial T}\right)_P = k_1\left(\frac{C^P}{C_V}\right)T^{(C_P/C_V)-1}$$

and the coefficient of isothermal compressibility,

$$\beta = -\frac{1}{V}\left(\frac{\partial V}{\partial P}\right)_T = \frac{k_2}{P}$$

In this problem, assume that C_P, C_V, K_1 and k_2 are constants.

PROBLEM 45. At high pressures and temperatures a quite good equation of state is $P(V - nb) = nRT$, where b is a constant. Find $(\partial V/\partial T)_p$ and $(\partial V/\partial P)_T$ from this equation and hence dV in terms of P, T, V-*nb*, *d*T and *d*P.

PROBLEM 46. (a) Estimate the Boyle temperature for a certain non-ideal gas whose equation of state is

$$P\overline{V} = RT + APT - BP,$$

where $\overline{V}$ is the molar volume and A and B are constants.
(b) Determine whether this gas has a critical point.

PROBLEM 47. Express the Boyle temperature of a van der Waals gas in terms of the constant a.b. and R. The final expression for T_B must contain only *a*, *b*, and R.

PROBLEM 48. Gases A, B, C, and D obey the van der Waals equation, with a and b values as given (liter-atm system of units):

	A	B	C	D
a	6	6	20	0.05
b	0.025	0.15	0.10	0.02

which gas has the highest critical temperature, the largest molecules, the most nearly ideal behavior at STP?

PROBLEM 49. The van der Waals constants for gases A, B, and C are:

Gas	a, liters^2 –atm/mole	b, liters/mole
A	4.0	0.027
B	12.0	0.030
C	6.0	0.032

which gas has (a) the highest critical temperature, (b) the largest molecular volume, and (c) most ideal general behavior around STP?

PROBLEM 50. Given the following data for a certain nonideal gas at 25°C:

p/P, g/liter-atm:	10	11	10	(P = density)

P, atm:	1	10	20

The critical pressure of this gas must then be (greater than 10 atm, greater than 20 atm, between 1 and 20 atm, between 1 and 10 atm, less than 20 atm, can't tell).

PROBLEM 51. Consider an equation of state in the form

$$P = \frac{RT}{V-b} \cdot e^{-A/RT^{3/2}\bar{V}}$$

for an arbitrary gas, Show that the critical constants are given by

$$T_c = \left(\frac{A}{4br}\right)^{2/3}$$

$$\bar{V}_c = 2b \text{ and } P_c = \frac{R}{b}\left(\frac{A}{4bR}\right)^{2/3} e^{-2}$$

PROBLEM 52. A gas obeys the van der Waals equation, with P_c = 30 atm and T_c = 200°C. The compressibility factor PV/RT will be more than one (at P = 50 atm, T = 250°C; at P = 1 atm, T = 100°C; P = 500 atm, T = 500°C: none of these). Calculate the van der Waals constant *b* for this gas.

PROBLEM 53. Derive the relation between the constants *a*, *b*, and R of the Berthelot equation,

$$P = \frac{RT}{(\bar{V}-b)} - \frac{a}{T\bar{V}^2}$$

and the critical parameters. Express the equation of state in terms of reduced variables.

PROBLEM 54. Some P–V plots are shown below for a gas that obeys the van der Waals equation. Calculate the constants a and *b* for this gas. Since your calculation is necessarily approximate, it is necessary to show very clearly just how you have obtained numbers from the graph below and how you have used them. Give the units of *a* and *b* also.

(*Hint*: The 300°C isotherm is closest to critical point).

PROBLEM 55. An equation of state due to Dieterici is

$$P\,(V - nb')\exp\frac{na'}{RTV} = nRT$$

Evaluate the constants a′ and b′ in terms of the critical constants P_c, V_c, and T_c of gas.

PROBLEM 56. The critical temperature and pressure for NO gas are 177°K and 64 atm, respectively, and for CCl_4 they are 550°k and 45 atm, respectively. Which gas (a) has the smaller value of the van der Waals constant *b*, (b) the smaller value of the van der Waals constant *a*, (c) is most nearly ideal in behavior at 300°K and 10 atm?

PROBLEM 57. A nonideal gas of molecular weight 150 obeys the van der Waals equation; its critical pressure and temperature are 100 atm and 100°C, respectively.

(a) The compressibility factor PV/RT will be greater than unity at (500 atm and 80°C, 500 atm and 120°C, 50 atm and 60°C, 50 atm and 120°C, none of these).

(b) Calculate the value of the compressibility factor at the critical point.

PROBLEM 58. Carbon dioxide is assumed to follow the equation of state

$$\left(P + \frac{n}{V^2\,T^{1/2}}\right)(v - m) = RT$$

where *n* and *m* are constants for any gas. Given that the critical pressure and temperature of carbon dioxide are 72.9 atm and 304.2°K respectively, determine the compressibility factor of the gas at 100°C and at a volume of 6.948 cubic decimeters per kilogram.

PROBLEM 59. A bubble with a diameter of 1 cm at the bottom of the Okeefenokee Swamp, where the temperature is 5°C and the pressure is 3 atm, rises to the surface, where the temperature is 25°C and the pressure is 1 atm. What will be the diameter of the bubble when it reaches the surface? (Ignore the presence of water vapor in the bubble.)

PROBLEM 60. The molecular weight ozone was determined by the Regnault method by Kerrer and Waulf(*J. Am. Chem. Soc.*, 44, 2391 (1922)], and the following data were obtained :

Temperature = 28.2°C

Weight of bulb filled with ozone = 6.7624 g

Weight of evacuated bulb = 6.5998 g

Pressure of ozone in bulb = 274.4 torr

Volume of bulb = 235.67 cm^3

Calculate the molecular weight of ozone from these data.

PROBLEM 61. MacInnes and Kreiling [*J. Am. Chem. Soc.*, 39, 2350 (1917)] determined the molecular weight of diethyl ether by the Victor Meyer method and obtained the following data :

Weight of ether = 0.1023 g

Volume of air displaced = 35.33 cm^3

Temperature = 32.5°C

Atmospheric pressure = 743.95 torr

Calculate the molecular weight of the ether

PROBLEM 62. A certain gas has the following densities at 300.0°K:

Pressure, atm	0.4000	0.8000	1.0000
Density, g liter^{-1}	1.512	3.088	3.900

Find the molecular weight of the gas as accurately as you can with the given data.

PROBLEM 63. The density of gaseous methylamine as a function of pressure at 0°C was investigated by Arthur and Felsing [*J. Am. Chem. Soc.*, 68, 1883 (1946)]. The following data were obtained :

Pressure, atm	0.200	0.500	0.800
Density, g liter^{-1}	0.2796	0.7080	1.1476

Determine the molecular weight of this compound.

PROBLEM 64. Using Maxwell's law of distribution of molecular energies, $dn = K\sqrt{e}\ \varepsilon^{-/kT}\ d\varepsilon$, write an expression for the fraction of all molecules in an ideal gas that have translational kinetic energies greater than a certain value ε_o

PROBLEM 65. Perrin [*Ann. Chim. Phys.*, 18, 1 (1909)] determined the value of Avogadro's number by investigating microscopically the distribution of colloidal gamboge particles suspended in water as a function of height. In one set of measurements he obtained the following data at 15°C:

h, height, *μm*	5	35
n, relative number of particles at height *h*, averaged over a period of time	100	47

The density of the particles was 1.206 g cm^{-3}. The density of the water was 0.999 g cm^{-3}. The radius of the particles was 0.212 μm. 1μm = 10^{-4} cm. On the basis of this information calculate Avogadro's number.

PROBLEM 66. Plot the probability density function *f*(ν) versus molecular speed ν for hydrogen at 0°C.

PROBLEM 67. Calculate the mean free path of argon molecules at 25°C and 1 torr. The molecular diameter of argon is 2.86 Å.

PROBLEM 68. Calculate the collision frequency of nitrogen molecules per cubic centimeter per second at 1 torr and 25°C. The molecular diameter of N_2 is 3.16 Å.

PROBLEM 69. A mixture consisting of 90 mole per cent hydrogen and 10 mole per cent deuterium at 25°C and a total pressure of 1 atm is permitted to effuse through a small orifice of area 0.30 mm^2. Calculate the composition of the initial gas that passes through.

PROBLEM 70. Calculate the pressure in atmospheres exerted by 2.00 moles of chlorobenzene vapor confined to 10.0 liters at 25°C (*a*) using the ideal gas law, and (*b*) using the van der Waals equation. $a = 25.43$ liter^2 atm mole^{-2}, $b = 0.1453$ liter mole^{-1}.

PROBLEM 71. The compressibility factor $Z = PV/nRT$ for CO_2 at 0°C and 100 atm pressure is 0.2007. Calculate the volume occupied by 0.1 mole of the gas at 100 atm and 0°C (*a*) by the ideal gas law, and (*b*) by making use of the compressibility factor.

PROBLEM 72. The compressibility factor $Z = PV/nRT$ for N_2 at –50°C and 800 atm pressure is 1.95; at 100°C and 200 atm it is 1.10. A certain mass of nitrogen occupied a volume of 1.00 liter at –50°C and 800 atm. Calculate the volume occupied by the same quantity of nitrogen at 100°C and 200 atm.

PROBLEM 73. For a van der Waals gas $C_c = 3b$, $P_c = a/27b^2$, and $T_c = 8a/27bR$. Evaluate numerically the compressibility factor of a van der Waals gas at the critical point.

PROBLEM 74. The boiling point of *n*-hexane at 760 torr is 68.9°C. Make a rough estimate of the critical temperature of this substance.

PROBLEM 75. The coefficient of thermal expansion of helium as a function of pressure is given below [Henning and Heuse, *Z, Physik*, 5, 285 (1921)]. Calculate the temperature in degrees Celsius corresponding to absolute zero.

P, torr	504.8	520.5	760.1	1102.9	1116.5
$\alpha \times 10^6$	3658.9	3660.3	3659.1	3658.2	3658.1

PROBLEM 76. The density ρ (in g liter^{-1}) of a certain gas at 300.00°K is given by the equation $\rho = 2.000P + 0.0200\,P^2$, where P is the pressure in atmospheres. Calculate the molecular weight of this gas to four significant figures.

PROBLEM 77. Gray and Burt [*J. Chem. Soc.*, 95, 1633 (1909)] determined the atomic weight of chlorine by the method of limiting densities applied to gaseous hydrogen chloride. In one set of measurements they found that a glass bulb with a volume of 465.856 cm^2 contained the following weights of hydrogen chloride at the indicated pressures and 0°C:

Weight of HCl, g	0.760.97	0.75731	0.75481	0.75968
Barometric pressure, mm Hg	756.76	753.38	750.83	755.79

In another set of experiments they obtained the following data on the PV product of hydrogen chloride as a function of pressure at 0°C :

P, mm Hg	829.50	604.71	384.13
V, cm^3	66.012	90.815	143.259
PV	54.757	54917	55030
P	243.40	157.57	
V	226.401	349.978	
PV	55106	55146	

They also found that the limiting density-pressure ratio of oxygen at 0°C was limp→0 (ρ/P) = 1.42762 g $liter^{-1}$ atm^{-1}. The measurements were made in London, where the force of gravity is 1.000588 times that at 45° latitude. All of the data have been corrected for the adsorption of gas on the walls of the container. Determine the limiting density of hydrogen chloride and the atomic weight of chlorine.

PROBLEM 78. The thermal coefficient of cubic expansion of mercury is given by the expression

$$\alpha = \frac{1}{V_0}\left(\frac{\partial V}{\partial T}\right)_P = 1.817 \times 10^{-4} + 5.90 \times 10^{-9} t + 3.45 \times 10\text{–}10t^2$$

where t is the Celsius temperature and V_0 is V at t = 0. If an ideal gas thermometer and a mercury thermometer are set to agree at 0°C and 100°C, what apparent temperature on the mercury scale would correspond to 50°C on the ideal gas thermometer?

PROBLEM 79. Each of the following definitions of absolute temperature (T) is possible and workable. Describe briefly how each definition differs from the official definition (adopted in 1954 by the International Committee on Weights and Measures), and give one or more reasons why you think each of these definitions is not commonly used. The official definition is:

$$T = \frac{1}{nR} \lim_{P \to 0} (PV)$$

where n is the number of moles of gas, V is its volume, P is its pressure, and R is a constant chosen so that $T = 273.16°$ (exactly) at the triple point of water.

$$(a)\ T = 100 \left(\frac{l - l_f}{l_b - l_b} \right) + 273.15$$

Where l is the length of a mercury column of constant cross section at the temperature T, and(l_b, l_t) is the length of the same column at the (boiling, freezing) point of air-saturated water under a pressure of one standard atm (1.01325×10^6 dyne cm^{-2}).

(*b*) T = PV/*n*R, where *V* is the volume at one atm pressure of *n* moles of pure oxygen of the isotopic composition found in the earth's atmosphere. One mole is defined as that quantity of oxygen (approximately 31.9988 g) which contains the same number of molecules as there are atoms in exactly 12 g of the pure nuclide ^{12}C. R is a constant so chosen that T = 273.16 at the triple point of water.

PROBLEM 80. A certain mixture of helium and argon weighing 5.00 g occupies a volume of 10.0 liters at 25°C and 1.00 atm. What is the composition of the mixture in weight per cent?

PROBLEM 81. Calculate the density in g $liter^{-1}$ of air saturated with water vapor at 25°C. The vapor pressure of water at this temperature is 23.7 torr. The composition of dry, CO_2 -free air is 78.1 per cent N_2, 21.0 per cent O_2, and 0.9 per cent Ar by volume.

PROBLEM 82. The pressure of a gas as function of height is given by the formula

$$\ln\left(\frac{P_0}{P}\right) = \frac{Mg(h - h_0)}{RT}$$

where P_0 represents the pressure of the gas at height h_0, and P represents the pressure of the gas at height h, $h > h_0$; M is the molecular weight of the gas, g is the gravitational acceleration constant = 980.6 cm sec^{-2}, and R is the gas constant = 8.314×10^7 ergs °K^{-1} mole^{-1} (*a*) Calculate the atmospheric pressure at the top of the Empire State Building (height 1248 ft) if the pressure at ground level is 1.00 atm and the temperature is 25°C, assuming no complications such as turbulence or temperature gradients. (*b*) Explain how you could construct a straight-line graph relating altitude to air pressure. The apparent molecular weight of air is 29.

PROBLEM 83. Derive the barometric formula

$$P = P_0\, e^{-Mgz/RT}$$

by solving a differential equation obtained as follows: The pressure difference ($-dP$) between height z and height $z + dz$ is equal to the weight (not mass) per unit area of a layer of gas of thickness dz. Then equate these two, express everything (by means of the ideal gas law) in terms of the two variables P and z and separate these variables.

PROBLEM 84. The acceleration of gravity, g, varies in magnitude with distance r from the center of the earth, in accordance with the formula $g = Gm_E/r^2$, where G is a proportionality constant called the gravitational constant and m_E is the mass of the earth. Derive a modified form of the barometric equation that takes this variation of g with altitude z above the earth's surface into account.

PROBLEM 85. In the lower part of the atmosphere the temperature of the air is not uniform but decreases linearly with altitude in accordance with the equation $T = t_0 - az$, where a is a proportionality constant, z is the altitude, T_0 is the temperature at ground level, and T is the temperature at altitude z. Derive a modified form of the barometric equation that takes this temperature dependence into account.

PROBLEM 86. Your answers to this question should be expressed in terms of the given quantities and universal constants. They may contain integrals that you have not evaluated. (*a*) In an ideal gas at the temperature T in which each molecule has mass m find the fraction of the molecules that have speeds greater than υ_0, (*b*) Find this fraction in a two-dimensional ideal " gas" with the molecules confined to a plane.

PROBLEM 87. The external surface area of an average human being is estimated to be about 2.0m^2. Calculate the number of collisions per second of air molecules (20 mole percent O_2, 80 mole per cent N_2) with the total surface of an average person at 25°C and 1 atm.

PROBLEM 88. The vapor pressure of scandium was determined by the Knudsen effusion technique [Krikorian, *J. Phys. Chem.*, 67, 1586 (1963)]. In this method the weight of gas effusing through a small opening in a known interval of time is measured. The following data were obtained:

temperature, 1555.4°K

time, 110.5 min

weight loss, 9.57 mg

diameter of orifice, 0.2965 cm

Calculate the vapor pressure of scandium at this temperature in atmospheres. 1 dyne cm^{-2} = 9.87×10^{-7} atm.

PROBLEM 89. Two ideal gases, A and B, are in separate containers at temperatures T_A and T_B, with concentrations c_A and c_B in moles per liter, respectively. Their molecular weights are M_A and M_B; the molecular diameters are σ_A and σ_B. We assume that the molecules are hard spheres. For each of the following quantities find the ratio of its value for gas A to its value for gas B: (*a*) average kinetic energy per molecule, (*b*) average speed of a molecule, (*c*) mean free path, and (*d*) number of intermolecular collisions suffered by a single molecule per unit time.

PROBLEM 90. According to the kinetic theory of gases the viscosity of a gas is proportional to $T^{1/2}\sigma^{-2}$, where σ is the molecular diameter. Explain why the viscosity (*a*) increases with increasing temperature, (*b*) depends on the diameter of the molecules, and (*c*) is independent of pressure at constant temperature. Why does this independence fail at very low pressures?

PROBLEM 91. It has been proposed that intergalactic space contains hydrogen atoms at a temperature of 7.8×10^5°K, with about 1 atom per 100 liters. Assume that the collision diameter of a hydrogen atom is 2.0Å. (*a*) Calculate the mean free path of an atom in intergalarctic space in light-years. (*b*) What is the approximate average time, in years, between collisions of a given atom?

PROBLEM 92. The Boyle temperature of a gas is the temperature at which $[\partial(PV)\partial P]T = 0$ when $P = 0$. For each of the following equations of state evaluate T_B, the Boyle temperature, in terms of the known constants A, b, R, P_c, T_c, etc.:

(a) $$P\overline{V} = RT + \left(b - \frac{A}{RT^{3/2}}\right)P$$

(b) $$P\overline{V} = RT\left\{1 + \frac{9}{128}\left(\frac{P}{P_c}\right)\left(\frac{T_c}{T}\right)\left[1 - 6\left(\frac{T_c}{T}\right)^2\right]\right\}$$

PROBLEM 93. Evaluate the Boyle temperature of a van der Waals gas in terms of the constants a, b, and R. The final expression for T_B should contain only a, b and R. Suggestion ;Since the van der Waals equation is difficult to rearrange into the viral form, which would permit T_B to be evaluated by inspection, set up the van der Waals equation in the form

$$P = \frac{nRT}{V - nb} - \frac{n^2 a}{V^2}$$

multiply through by *V*, and set

$$\left(\frac{\partial [PV]}{\partial TP}\right)_r = 0$$

at $P = 0$. Remember that when $P = 0$, $V = \infty$.

PROBLEM 94. Show that at low densities the van der Waals equation

$$\left(P + \frac{a}{V^2}\right)(V - b) = RT$$

and the Dieterici equation

$$P = \frac{RT}{(\overline{V} - b)} e^{-a/RTP}$$

give essentialy the same result for *P*. Hint: Expand the Dieterici equation in terms of the infinite series $e^x = 1 + x + x^2/2! + x^3/3! + x^4/4! + ...$

PROBLEM 95. A certain gas has the equation of state

$$P = \frac{RT}{\overline{V} - bT} - \frac{a}{\overline{V}^3}$$

where *V* represents molar volume, and *a, b* are constants characteristics of the gas. Show that

$$a = \frac{4RT_c \overline{V}_c^2}{3} \text{ and } b = \frac{\overline{V}_c}{2T_c}$$

where $\overline{V}_c$ and T_c are the critical constants.

PROBLEM 96. A certain nonideal gas has the equation of state $P\overline{V} = RT + APT - BP$, where $\overline{V}$ is the molar volume and *A* and *B* are constants characteristic of the gas. For this gas (*a*) find the Boyle temperature, and (*b*) ascertain whether this gas has a critical point. If so, find the critical constants: if not, explain why not.

PROBLEM 97. A hypothetical gas the equation of state

$$P = \frac{RT}{\overline{V} - b} - \frac{a}{\overline{V}}$$

where a and b are constants distinct from zero. Ascertain whether this gas a critical point. If it has, express the critical constants in terms of a and b. If it has not, explain how you arrive at this conclusion.

PROBLEM 98. For a Dieterici gas, $P(V - b)e^{a/RTV} = RT$; $\overline{V}_e = 2b$, $T_c = a/4Rb$, and $P_c = a/4e^2b^2$. Rewrite the Dieterici equation in reduced from –that is, in terms of reduced temperature, $\theta = T/T_e$, reduced pressure, $\pi = P/P_e$, reduced volume, $\phi = \overline{V}/Vc$. In the final equation π, ϕ, and θ should be the only variables and the constants a, b, and R should not appear at all.

PROBLEM 99. In filling a barometer tube some air was trapped at the top of the column, as was evident from the fact that the instrument read 3.0 mm low when the true pressure was 720 torr (mm Hg) and the room temperature was 20°C. The total length of the barometer tube measured from lower index was 780 mm. Work out a formula from which the true pressure could b determined with this instrument under all room conditions. Solve for the correction to be applied to the reading.

PROBLEM 100. A glass cylinder 100 cm long, closed at one end, and filled with dry air at 1.00 atm and 25°C is inverted in a pool of mercury until the closed end is flush with the surface of the pool (see the figure 2 below). The barometric pressure is 1.00 atm. (*a*) What will be the height of the mercury column inside the cylinder? (*b*) What will be the final pressure of the trapped air?

PROBLEM 101. The J-shaped tube shown in the figure is filled initially with air at 1.00 atm and 25°C; then liquid mercury is poured into the open end. The barometric pressure is 1.00 atm. (*a*) What will be the height of the mercury column in the long arm when mercury is just starting to spill over the top of the

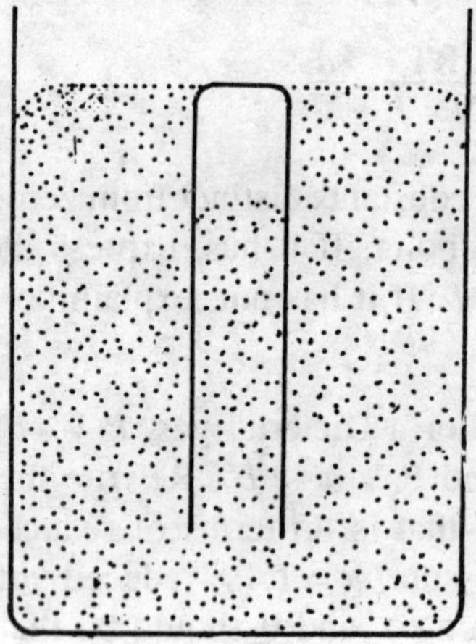

Fig. 2.

short arm? (*b*) What will be the pressure of the trapped air at this point?

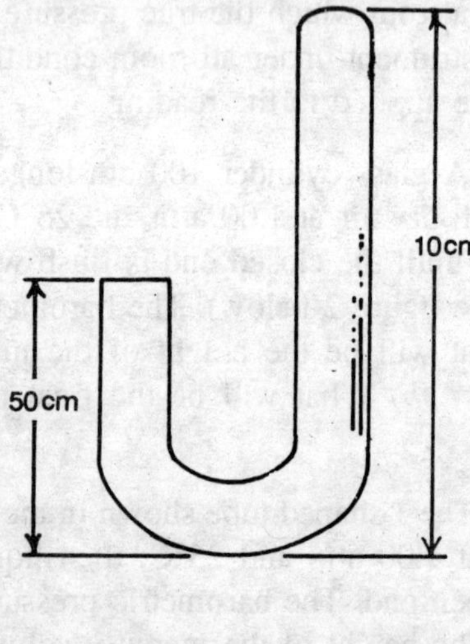

Fig. 3.

PROBLEM 102. Two hollow, spherical vessels with a capacity of 22.4 liters each are each filled with 1.00 mole of nitrogen at 25°C and 1.00 atm pressure and connected by a thin, hollow tube of negligible volume. Then one sphere is placed in a thermostat at 100°C and the other sphere is placed is a thermostate at 0°C. What will be the final pressure in the system and what will be the final number of moles of nitrogen in each sphere?

PROBLEM 103. In an attempt to derive the formula PV = KT, where *K* is a constant, by combining the laws of Boyle and Charles (or Gay-Lussac) the following incorrect argument was proposed:

$PV = k_1$	(1)	$V = k_2 T$	(4)
$\frac{V}{T} = k_2$	(2)	$\frac{k_1}{P} = k_2 T$	(5)
$V = \frac{k_1}{P}$	(3)	$\frac{k_1}{k_2} = k_3 = PT$, or $PT = K$	(6)

Equation 6 is obviously incorrect. What is the error? At what step was it introduced? Give a correct derivation.

PROBLEM 104. A chemical engineer wishes to construct a simple, straight-line graph by means of which he could read the composition of a mixture of helium and oxygen in mole percent off one axis by looking up the density, in grams per liter, at 25°C and 1 atm total pressure on the other axis, assuming ideal gas behavior.

(*a*) Explain the basis on which such a graph can be constructed.

(*b*) Draw the graph and use it to determine the composition of a helium-oxygen mixture that has a density of 0.820g liter^{-1} under the specified conditions.

PROBLEM 105. The three components of the velocity of a gas molecule are υx, υy and υz. The speed is $\upsilon = (\upsilon_{x2} + \upsilon_{y2} + \upsilon_{z2})_{1/2}$.

(*a*) Write an expression for υ_{x4}, the fourth power of υ_x averaged over all molecules. (This and later answers may be left in the form of unevaluated integrals. However, they should contain only the mass m of each molecule, the temperature T, universal constants, and the variable(s) of integration.)

(*b*) Write a similar expression for υ^4.

(*c*) If there is a difference between your two expressions, other than the substitution of υ for υ_x, explain how it arises.

PROBLEM 106. Sketch a representative isotherm for 1 mole of a gas that obeys the equation of state $P(\overline{V} - b) = RT$, where $b > 0$. Interpret the physical significance of the slope and the PV intercept and discuss the validity of this equation of state for real gases at low pressures.

PROBLEM 107. For a certain hypothetical gas the thermal coefficient of expansion is given by

$$\alpha = \frac{1}{V}\left(\frac{\partial V}{\partial T}\right)_P = k_1\left(\frac{C_P}{C_v}\right)T^{(C_P/C_V)-1}$$

and the coefficient of isothermal compressibility by

$$\beta = -\frac{1}{V}\left(\frac{\partial V}{\partial P}\right)_T = \frac{k_2}{P}$$

What is the equation of state of this gas? Assume that C_p, C_v, k_1, and k_2 are constants.

PROBLEM 108. For a van der Waals gas show that $T_e = 8a/27bR$, $V_c = 3nb$, and $P_c = a/27b^3$, where T_c is the critical temperature, V_c is the critical volume, and P_c is the critical pressure. Van der Waals' equation is

$$\left(P + \frac{n^2 a}{V^2}\right)(V - nb) = nRT$$

PROBLEM 109. A gas has the equation of state $(P + aP^2)$ $(V - b) = RT$, where a and b are constants. Ascertain whether this gas has a critical point. If it has, write the equation of state in reduced form.

PROBLEM 110. For a gas described by the Berthelot equation,

$$P = RT/(\overline{V} - b) - a/T\overline{V}^2,$$

express *a, b,* and *R* in terms of the critical constants and write the equation of state in terms of reduced variables.

PROBLEM 111. Show that the critical constants T_c, V_c, and P_c for a gas with the equation of state

$$P = \frac{RT}{1-b} \cdot e^{A/RT^{3/2]}P}$$

are given by

$$T_c = \left(\frac{A}{4bR}\right)^{2/3}, \ \overline{V}_c = 2b \text{ and } \quad P_c = \frac{R}{b}\left(\frac{A}{4bR}\right)^{2/3} e^{-2}$$

PROBLEM 112. The following equation of state for a nonideal gas is proposed :

$$P = \frac{RT}{\overline{V} - B} - A\,\overline{V}^m$$

where $\overline{V}$ is the molar volume, and *A, B,* and *m* are constants characteristic of the gas. If this equation is to be applicable to a real gas, it must predict (i) the approach to ideal behavior at zero pressure or infinite volume—that is, $PV/RT \to 1$ as $\overline{V} \to \infty$; (*ii*) the existence of a critical point, with V_c, T_c, and P_c all positive. We wish to ascertain what conditions must be imposed on the constant *A, B,* and m is order that the equation will make these two predictions. (*a*) Express $P\overline{V}/RT$ in terms of *A, B, m, R, T,* and *V*. (*b*) If *A* and $B \neq 0$, what requirement(s) must be imposed in order that $PV/RT \to 1$ as $1 \to \infty$? (*c*) Express $(\partial P/\partial \overline{V})_T$ and $(\partial^2 P/\partial \overline{V}^2)_T$ in terms of *A, B, m* R, T, and *R*. (*d*) Express the critical volume V_c in terms of *A, B, m* and *R*. (*e*) What requirement(s) must be imposed on *A, B,* or *m* in order that $\overline{V}_c$ be positive? (*f*) Express the critical temperature T_c in terms of *A, B, m,* and *R*. (*g*) What further condition(s) must be imposed on *A, B,* or *m* in order that T_c be positive? (*h*) Discuss the physical significance of the conditions which *A, B* and *m* must satisfy.

Part-II
SOLUTIONS TO PROBLEMS

GASES AND KINETIC THEORY

Solution:1. (a) $\frac{F-32}{9}=\frac{C}{5}$

where F=Fahrenhite scale

C=Celsius or Centigrade scale

The problem states that the celsius scale is -10°C. Therefore, using equation (1),

$$\frac{F-32}{9}=\frac{-10}{5}$$

or

$$5(F-32)=9(-10)$$

$$5F-160=-90$$

$$5F=70$$

$$F=70/5=14$$

(b) 0° C is equivalent to 273 °K where K=absolute Kelvin scale. To convert C to K, use the relationship K= 273 + C. The problems above indicates that K =-10 therefore K=273 + (-10) =273-10=263.

Solution 2. K=273 + C. First, convert 77 Fahrenheit (F) to-Celcius (°C). Then convert Celsius (°C)to the absolute temperature scale, Kelvin (°K).Using $\frac{F-32}{9}=\frac{C}{5}$, substitute the value of F to obtain

$$\frac{77-32}{9}=\frac{C}{5}$$

$$5(77-32)=9C$$

$$5(45)=9C$$

$$C=25$$

K=273+ C

K=273+ 25= 298

Solution 3. This is a matter of scale conversions. In a problem dealing with pressure, it is reasonable to use a scale of one Atmospheric pressure = 1.013 10^5 N/m^2 =760mm Hg for conversions. Note the convenience in appropriate scale choice.If an inappropriate scale had been chosen, another conversion to someother scale would be needed before connverting to the desired scale. Since 1.013 × 10^5 N/m_2 is equivalent to 760 mm Hg, therefore 1.00 × 10^5 N/m^2 will be equivalent

$$\frac{1.00 \times 10^5 \text{ N/m}^2}{1.013 \times 10^5 \text{ N/m}^2} \times 760 \text{ mm Hg} = 750 \text{ mm Hg}.$$

Solution 4. The gallium column can be computed by the use of the definition:

(h_{Ga}) (d_{Ga}) =(h_{Hg}) (d_{Hg})

where h=height of a column,

d=density,

therefore $$h_{Ga} = \frac{(h_{Hg})\,(d_{Hg})}{(d_{Ga})}$$

1 atm pressure = h_{Hg} = 760 mm Hg= 76.0 cm Hg. The density of mercury =d_{Hg} = 13.6 gm/cm^3 . Therefore,

$$h_{Ga} = \frac{(76.0 \text{ cm Hg})\,(13.6 \text{ gm/cm}^3)}{(6.09 \text{ gm/cm}^3)}$$

$$= 169.7 \text{ cm Ga} = 1697 \text{ mm Ga}$$

Solution 5. The height of the mercury column will correspond to the pressure of the atmosphere. Now by definition

$$(h_{water})\,(d_{water}) = (h_{mercury})\,(d_{mercury})$$

where

d = density

h = height.

From the mathematical relationship given,

(600mm) (1 × 10^3 kg/m^3)=$(h_{mercury})$ 13.6×10^3 Kg/m^3)

$$h_{mercury} = 600 \text{ mm} \left[\frac{1 \times 10^3 \; kg/m^3}{13.6 \times 10^3 \; kg/m^3}\right]$$

or

$$h_{mercury} = \left[\frac{600}{13.6}\right] \text{mm Hg} = 44.1 \text{ mm Hg}.$$

Solution 6. To obtain the pressure of the gas in mm Hg, it is necessary to find the P_{Hg} that is equivalent to 172 mm of the liquid of specific gravity 1.23. By definition,

$$pH_g \; dH_g = p_{liq} \; d_{liq}$$

where pH_g = pressure of mercury

P_{liq} = pressure of liquid

d_{hg} = density of mercury

d_{liq} = density of liquid.

Therefore

$$PH_g = \frac{(P_{liq})\,(d_{liq})}{(dH_g)}$$

$$dH_g = 13.6 \text{ gm/cm}^3$$

$$d_{liq} = 1.23 \text{ gm/cm}^3$$

Here the density of the liquid is the same as the specific gravity because the specific gravity is given by

$$\frac{\text{density of substance}}{\text{density of water at 4°C}}$$

Density of liquid = (specific gravity) (density of water at 4°C).

But the density of water at 4°C = 1.00 grams/cm^3.

Density of liquid = specific gravity.

Therefore the specific gravity is numerically equal to density in the metric system. Specific gravity is dimensionless (has no units) whereas density has units of gm/cm^3 in the metric system. From the mathematical relationship,

$$PH_g = \frac{(172\text{ mm})(1.23\text{ gm/cm}^3)}{(13.6\text{ gm/cm}^3)} = 15.36\text{ mm Hg.}$$

Since atmospheric pressure supports the gas pressure and height of liquid, the following relationship is true:

$$Pa = Pg + PHg.$$

Remember that P_{Hg} is the pressure equivalent to the height of liquid.

$$Pa = 763\text{ mm Hg}$$

$$PH_g = 15.35\text{ mm Hg}$$

$$Pg = 763 - 15.55 = 747.45\text{ mm Hg.}$$

Solution 7. Always equate pressures at the lower liquid surface, that is, $P_g = P_a + P_{Hg}$. The gas pressure is greater than atmospheric pressure as shown in Fig. 1. The diagram shows the gas supporting both the mercury and the atmosphere.

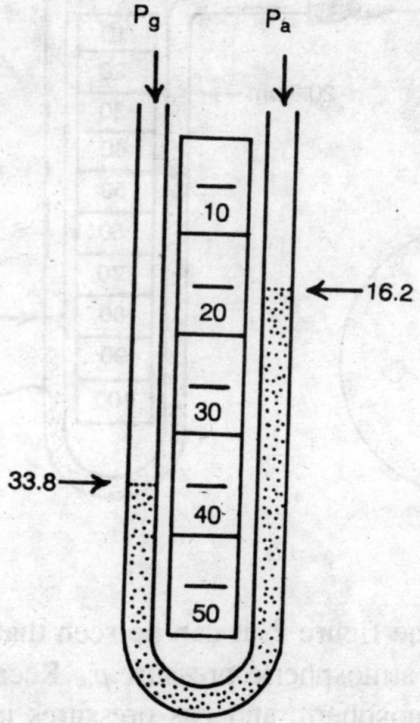

Fig.

The difference between atmospheric and gas pressure is always equal to the difference in levels of the mercury in the two legs of the manometer. This difference is 33.8-16.2 = 17.6, or 176 mm. Since the gas pressure is greater than atmospheric pressure,

$$Pg - Pa = 176 \text{ mm Hg}$$
$$Pa = 747 \text{mm Hg}$$
$$Pg = 747 + 176.0$$
$$= 923 \text{ mm Hg}.$$

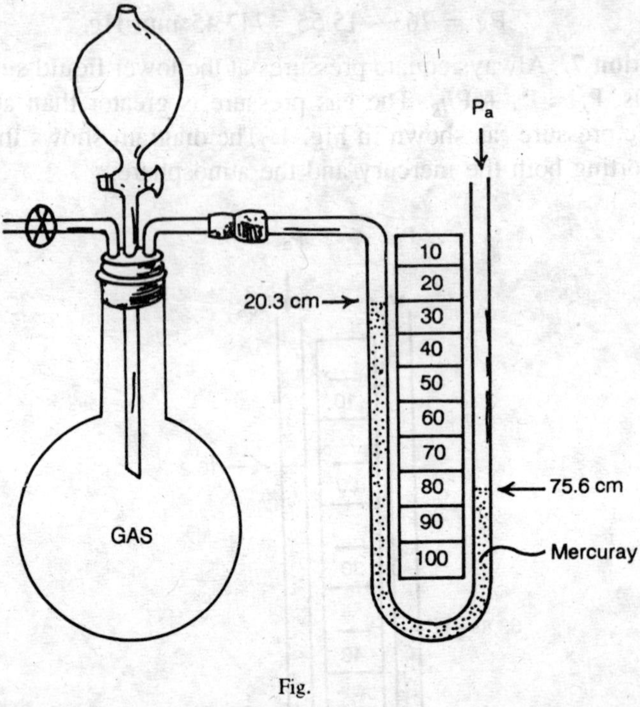

Fig.

Solution 8. From the figure 2, it can be seen that, the gas pressure Pg is less than atmospheric pressure p_a. Recall that the difference between atmospheric and gas pressures is always equal to the difference in levels of the mercury in the two legs of the manometer. This difference is 75.6 – 20.3 = 55.3 cm or 553 mm.

Then one can write,

$$Pa - Pg = 553 \text{ mm Hg}$$

since $$Pa = 755 \text{ mm Hg}$$

$$P_g = 755 - 553 = 202 \text{ mm Hg.}$$

Solution 9. The rate of molecular flow or efusion in molecules per-second is given by , dn/dt . Thus rate of flow of

$$H_2 = \frac{d n_{H_2}}{dt} \qquad (1)$$

where nH_2 is in $\frac{\text{molecules}}{\text{cm}^3} = {}^{n}H_2\, \bar{c}H_2\, A$ and $\bar{c}$ is the average speed in cm/sec and A is the area in cm^2. The rate of effussion of

$$D_2 = \frac{d n_{D_2}}{dt} = {}^{n}D_2\, \bar{c}D_2\, A \qquad (2)$$

The ratio of equation (1) to equation (2) is

$$= \frac{d n_{H_2}/dt}{d n_{D_2}/dt} = \frac{{}^{n}H_2\, \bar{c}H_2\, A}{{}^{n}D_2\, \bar{c}D_2\, A}$$

$$= \frac{{}^{n}H_2\, \bar{c}H_2}{{}^{n}D_2\, \bar{c}D_2}.$$

$\frac{{}^{n}H_2}{{}^{n}D_2} = \frac{\text{number of molecules of } H_2/\text{cm}^3}{\text{number of molecules of } D_2/\text{cm}^3}$ The volumes cancel in both the numerator and the denominator. To solve the problem molecules should be changed to moles.

$$\frac{\text{number of molecs of } H_2}{\text{number of moles of } D_2}$$

$$= \frac{\text{number of molecules of } H_2/\text{Avagadro \#}}{\text{number of molecules of } D_2/\text{Avagadro \#}}$$

So the ratio of mole per cent equals the ratio of molecular per-cent.

Therefore,

$$\frac{dn_{H_2}/dt}{dn_{D_2}/dt} = \frac{90}{10}\,\frac{\bar{c}_{H_2}}{\bar{c}_{D_2}} \quad (3)$$

But the rate of effusion is inversely proportional to the $m^{1/2}$ in *g*/mole. The root-mean-square velocities are proportional to arithmetical mean velocities, and since the proportionality constant cancels out equation (3) is written as

$$= \frac{dn_{H_2}/dt}{dn_{D_2}/dt} = \frac{90}{10}\sqrt{\frac{M_{D_2}}{M_{H_2}}}$$

$$= \frac{90}{10}\sqrt{\frac{4.028}{2.016}}$$

$$= 9\sqrt{2.0}$$

$$= 12.7$$

Thus 12.7 molecules of H_2 per molecule of D_2 effuses, and the per cent composition of the initial gas passing through will be

$$\text{mole \% } D_2 = \frac{1 \times 100}{12.7 + 1}$$

$$= 7.30\%$$

$$\text{mole \% } H_2 = 100 - 7.30$$

$$= 92.7\%$$

Solution 10. The total number of molecules striking the unit area of the wall of the cylinder is given by,

$$Z_{wall} = \frac{1}{4} n\bar{c}$$

$$= \frac{1}{4}\left(\frac{PN}{RT}\right)\bar{c}$$

$$= \frac{1}{4}\left(\frac{PN}{RT}\right)\left(\frac{8RT}{\pi M}\right)^{1/2}$$

$$Z_{wall} = \frac{1}{4}\frac{N}{RT}\left(\frac{8RT}{\pi}\right)^{1/2}\frac{P}{M^{1/2}}$$

N is the Avogadro's number and M = molecular weight. Thus it follows that at constant temperature the rate of effusion varies inversely as the square root of the molecular weight, M.

Therefore, the effusion of H_2 is proportional to $P_{H_2} m_{H_2}^{-1/2}$ and effusion of N_2 is proportional to $\{P_{N_2} M_{N_2}^{-1/2}$

$$\frac{z_{H_2}}{z_{N_2}} = \frac{P_{H_2}/M_{H_2}^{1/2}}{P_{N_2}/M_{N_2}^{1/2}}$$

$$\frac{z_{H_2}}{z_{N_2}} = \frac{P/(2)^{1/2}}{2P/(28)^{1/2}}$$

$$= \frac{1}{(2)^{1/2}}\left[\frac{(28)^{1/2}}{2}\right]$$

$$\frac{z_{H_2}}{z_{N_2}} = 1.87$$

or $$z_{H_2} = 1.87\ Z_{N_2}$$

Thus the rate of effusion of H_2 into N_2 compartment is 1.87 times faster than that of N_2 into the H_2 compartment of the cylinder.

Solution 11. (*a*) Assuming that the 1-dm^3 vessel is occupied by N particles, the molecular concentration is thus,

$$n = \frac{N}{V}$$

where V is the volume of the vessel. The total number of molecules striking the unit area of the wall, Zw, per second is given by:

$$Z_{wall} = \frac{1}{4} n \bar{c} \qquad (1)$$

where $\bar{c}$ is the mean speed of the molecules given by,

$$\bar{c} = \left(\frac{8RT}{\pi M}\right)^{1/2} \text{ and } n = \frac{N}{V}.$$

From equation (1),

$$Z_{wall} = \frac{1}{4}\frac{N}{V}\bar{c} \qquad (2)$$

But, N/V = PL/RT where P is the pressure and L is the Avogadro number. Therefore, equation (2) becomes

$$Z_{wall} = \frac{1}{4}\frac{PL}{RT}\bar{C}$$

Define dn/dt = moles striking the small hole of area A per sec.
= rate of change in moles of gas per sec.

$$-\frac{dn}{dt} = Z_{wall}\frac{A}{L}$$

$$= \frac{1}{4}\frac{PL}{RT}\bar{c} \times \frac{A}{L}$$

$$-\frac{dn}{dt} = \frac{PA\bar{c}}{4RT} \qquad (3)$$

From ideal gas law,

$$PV = nRT$$

or

$$n = PV/RT \qquad (4)$$

Differentiating equation (4) with respect to P and V,R,T, held constant gives,

$$dn = \frac{V\,dP}{RT}$$

substitute dn = VdP/RT into equation (3). This gives

$$= \frac{V}{RT}\frac{dP}{dT} = \frac{PA\bar{c}}{4RT}$$

or

$$-\frac{V}{RT}\frac{dP}{P} = \frac{A\bar{c}}{4RT\,dt}$$

$$-\frac{V}{RT}\int_{P_1}^{P_2}\frac{dP}{P} = \frac{A\bar{c}}{4RT}\int_0^t dt$$

$$-\int_{P_1}^{P_2}\frac{dP}{P} = \frac{A\bar{c}}{4V}\int_0^t dt$$

$$-\ln P\Big|_{P_1}^{P_2} = \frac{A\bar{c}}{4V}\,t$$

$$-\ln\frac{P_2}{P_1} = \frac{A\bar{c}}{4V}\,t$$

$$-\ln\frac{0.5}{1.0} = \frac{\pi r^2\bar{c}}{4V}\,t$$

$$0.693 = \frac{\left[\left(\frac{0.2\times10^{-6}\text{m}}{2}\right)\left(\frac{100\text{ cm}}{1\text{m}}\right)\right]^2\left(2.99\times10^4\text{cm sec}^{-1}\right)t}{4(10^3\text{ cm}^3)}$$

$$0.693 = 2.351\times10^{-9}\text{ sec}^{-1}t$$

$$t = \frac{0.693}{2.351\times10^{-9}\,sec^{-1}}$$

$$= 2.948\times10^8\text{ sec}$$

$$t = (2.948\times10^8\text{ sec})\left(\frac{1\text{ hr}}{3600\text{ sec}}\right)$$

$$t = 8.18\times10^4\text{ hr.}$$

(*b*) The assumption made in the calculation was that the mean free path λ of the gas was very much larger than the diameter of the hole. To justify this, look at the average number of collisions experienced by one molecule per second, given by

$$Z = 2^{1/2}\pi N d^2 \bar{c}$$

In this time, the molecule has traveled a distance $\bar{c}$. The mean free path λ is therefore $\bar{c}/Z$, or

$$\lambda = \frac{1}{2^{1/2}\pi N d^2} \qquad (5)$$

where N is the number of molecules = PL/RT and d is the diameter of hole. From (5) then

$$\lambda = \frac{1}{2^{1/2}\pi \frac{PL}{RT} d^2} \qquad (6)$$

$$\frac{PL}{RT} = \frac{(1/760 \text{ atm})(6.02 \times 10^{23})}{(82.06)(300)}$$

$$= 3.2 \times 10^{16}$$

and $\quad d = 3.70 \times 10^{-8}$ cm.

Substituting d and PL/RT into (6) gives

$$\lambda = \frac{1}{\sqrt{2}\,\pi\left(3.2 \times 10^{16}\right)\left(3.70 \times 10^{-8}\right)^2}$$

$$\lambda = 5.11 \times 10^{-3} \text{ cm.}$$

Thus λ is >> diameter of hole.

Solution 12. (a) Gas viscosities are determined by the rate at which momentum is transported between adjacent layers of fluid that are moving at different speeds. Viscosity increases with increasing temperature because the mean sped, $\overline{C}$, of the molecules is proportional to $T^{1/2}$. This arises because at higher temperatures momentum is transported more rapidly through a given area, and so force has to be increased to mintain the motion of the layers of gas.

Another way of looking at it is to consider that molecules

are not actually hard sphers, but somewhat soft and surrounded by force fields. Thus the higher the temprature, the faster the molecules move, and the deeper one molecule can penetrate into the force field of another, before repulsion.

(*b*) An increase in molecular diameter of the gas molecules results in frequent collisions with other molecules, and hence a shorter mean free path for the molecules.

This reduction in the mean free path of the molecules results in a reduction in the average momentum difference between the layers of the gas between which the molecules travel in each free path. Thus, viscosity decreases with increasing molecular diameter.

(*c*) The frequency of momentum – transporting molecules or molecules moving back and forth between layers increases in proportion to the pressure, but the mean free path, or effective distance between layers, decreases in proportion to the pressure. These two effects just cancel with respect to the momentum transport per unit distance.

The above explanation is true at high pressures. At very low pressures the mean free path does not increase further and so the viscosity approaches zero as the pressure approaches zero.

Solution 13. The Poiseuille's equation can be written as

$$d\frac{V}{dt} = \frac{(P_1^2 - P_2^2)\,\pi R^4}{16 l \eta\, P_0} \qquad (1)$$

or

$$\eta = \frac{(P_1^2 - P_2^2)\,\pi R^4}{16 l \left(d\frac{V}{dt}\right) P_0}$$

where V is the flowing volume , P_1 and P_2 are the pressures at each end of the tube of length l, and P_0 is the pressure at which the volume is measured. η is the coefficient of viscosity (or viscosity) and is defined as a measure of the friction present when adjacent layers of the gas are moving at different speeds. Evaluating $(P^2{}_1 - P^2{}_2)$ in equation (1) gives

$$(P_1^2 - P_2^2) = \frac{(765^2 - 760^2)\ \text{mmHg} \times (1.0135 \times 10^5\ \text{Nm}^{-2})^2}{(760\ \text{mm Hg})^2}$$

$$= 1.356\times 10^8\ \text{N}^2 m^{-4}$$

from the data of the problem set,

$$\frac{dV}{dt} = \left(\frac{90.2\ \text{cm}^3}{100\ \text{sec}}\right)\left(\frac{\text{lm}}{100\ \text{cm}}\right)^3$$

$$= 9.02 \times 10{-}7\ m^3\ s^{-1}$$

Thus from (1)

$$\eta = \frac{1.356 \times 10^8 \text{N}^2 m^{-4})\ (\pi)\ (1 \times 10^{-3} m)^4}{16(\text{lm})\ (1.0135 \times 10^5 \text{N}m^{-2})\ (9.02 \times 10^{-7} m^3 s^{-1})}$$

$$= \frac{0.267 \times 10^8 \times 10^{-12}}{146.27 \times 10^{-7} \times 10^5} = \frac{0.267 \times 10^{-4}}{146.27 \times 10^{-2}}$$

$$= 0.00182 \times 10^{-2}$$

$$\eta = 1.82 \times 10^{-5}\ \text{kg m}^{-1}\ \text{s}^{-1}$$

Solution 14. In the Knudsen method, the weight of gas effusing through a small orifice in a given time interval is monitored and this is proportional to the vapor pressure, P.

$$P = Z\left(\frac{2\pi RT}{M}\right)^{1/2} \qquad (2)$$

where Z is the collision frequency and is given by

$$Z = \frac{\Delta m}{A\Delta t} = \frac{\Delta m}{\pi r^2 \Delta t}$$

where

A = area of container

m = weight loss .

Therefore,

$$Z = \frac{(9.57\ mg)\left(\frac{1g}{1000\ mg}\right)}{\pi\ (0.2965/2)^2\ (110.5\ \text{min})\left(\frac{60\ \text{sec}}{1\ \text{min}}\right)}$$

$$= 2.08 \times 10^{-5} g\ \text{cm}^{-2\ \text{sec}}{-1}$$

Using this value in equation (1) gives

$$P = 2.08 \times 10^{-5}\ g\ cm^{-2}\ \text{sec}^{-1}$$

$$\left[\frac{2\pi(8.314 \times 10^7\ \text{ergs}\ °K^{-1} - \text{mole}^{-1})\ (1555.4°K}{44.96\ g/\text{mole}}\right]^{1/2}$$

$$= 2.80\ \text{dynes}\ cm^{-2}$$

$$P = (2.80\ \text{dynes cm}^{-2})\left(\frac{9.87 \times 10^{-7}\ \text{atm}}{1\ \text{dyne}\ cm^{-2}}\right)$$

$$P = 2.77 \times 10^{-6}\ \text{atm}.$$

Solution 15. In the method of Knudsen, if a solid has a vapor pressure, P, and is enclosed in a box with a small hole, the rate of loss of mass from the box is proportional to P. Therefore P may be determined if the mass loss is known. Let the total mass diffused = M_T .

Inside the container, the number of collision per unit time per unit area is

$$Z = \frac{1}{4}\left(\frac{\overline{C}NM}{L}\right) \quad (1)$$

where Z is the collision frequency and $\overline{C}$ is the mean speed of molecules given by

$$\overline{C} = \left(8\frac{RT}{\pi M}\right)^{1/2}$$

But $P = NRT/L$ (2)

Rearranging equation (1) to solve for L gives

$$L = \left(\frac{\overline{C}NM}{4Z}\right)$$

Inserting this in equation (2) yields

$$P = \frac{NRT}{\left(\frac{\bar{C}NM}{4Z}\right)} = \frac{NRT}{1} \times \frac{4Z}{\bar{C}NM}$$

$$P = \frac{Z4RT}{M\bar{C}} \tag{3}$$

Moreover, the definition of Z indicates that

$$Z = \frac{M_T}{tA} = \frac{M_T}{t\pi r^2}$$

Hence, equation (3) becomes

$$P = \frac{M_T}{tr^2}\left(\frac{2RT}{\pi M}\right)^{1/2}$$

$$= \frac{(0.00888)}{15.2 \times 60(0.159)^2}\left(\frac{2.8314 \times 10^7 \times 1537}{3.14 \times 9.012}\right)^{1/2}$$

$$= 36.60 \text{ dynes cm}^{-2}$$
$$= 3.61 \times 10^{-5} \ atm.$$

Solution 16. Assume that the atoms impinging on the wall of the oven are doing so from the *x*-direction, so only their velocity components in that direction need be considered. The velocity of the atoms are thus between zero and infinity . The mean velocity $<V_x>$, of a participle is given by

$$<V_x> = \int_0^{\infty} V_x f(V_x)\, dV_x \tag{1}$$

where $f(V_x)$ is the velocity distribution gives as

$$f(V_x) = (m/2\pi kT)^{1/2} \exp(-1/2\ mV_x^2/kT) \tag{2}$$

Also $\int_{-\infty}^{\infty} f(V_x)\, dV_x = 1 = 1$

The velocity component lies in the range $-\infty \le V_x \le +\infty$; hence, its probability must be unity.

In order to make the integral $\int_{o}^{\infty} f(V_x)\, dV_x = 1$, it is necessary to multiply the right-hand side of equation (2) by 2. Therefore from equation (1)

$$<V_x> = 2(m/2\pi kT)^{1/2} \int_{o}^{\infty} V_x \exp(-mV_x^2/2kT)dV_x$$

$$= 2(m/2\pi kT)^{1/2}\,(kT/m) = \sqrt{(2kT/\pi m)}$$

For Caesium at 500°C,

$$<V_x> = \left[\frac{(2/\pi)\times(1.381\times 10^{-23}\ JK^{-1})\times(773\ K)}{(132.9)\times(1.6605\times 10^{-27}\ kg)}\right]^{1/2}$$

$$<V_x> = 175.5\ m - s^{-1}$$

Solution 17. The root mean square velocity is defined by

$$U_{rms} = \sqrt{\overline{U^2}} \qquad (1)$$

where $\overline{U^2}$ = average of the square of the velocity

U_{rms} = root mean square velocity.

From the P–$\overline{V}$ – KE relationship,

$$P\overline{V} = \tfrac{1}{3} M\overline{U^2} = \frac{2}{3} KE \qquad (2)$$

where P = Pressure

$\overline{V}$= molar volume

KE = Kinetic energy

M = molecular weight

But $P\overline{v}$ = RT . Therefore equation (2) becomes

$$RT = \frac{1}{3} M\overline{u^2} \qquad (3)$$

where R = universal gas constant

T = temprature.

Rearranging equation (3) to solve for $\overline{U^2}$ gives

$$\overline{U^2} = \frac{3RT}{M} \tag{4}$$

Substituting equation (4) into equation (1)gives

$$U_{rms} = \sqrt{\frac{3RT}{M}} = \left(\frac{3RT}{M}\right)^{1/2} \tag{5}$$

Squaring both sides of equation (5) gives

$$U^2_{rms} = \frac{3RT}{M}$$

Solving for T, the equation becomes

$$T = \frac{MU^3_{rms}}{3R} \tag{6}$$

From the given data,

$$T = \frac{(2\ g\ mol^{-1)}\ (1.1 \times 10^{6)^2}\ cm^2\ sec^{-2}}{(3)\ (8.3 \times 10^7\ ergs\ mol^{-1}\ deg^{-1})}$$

$$T = \frac{2.42 \times 10^{12}\ g\ cm^2\ sec^{-2}}{24.9 \times 10^7\ ergs\ deg^{-1}}$$

The units above might seem strange but erg is an energy unit in the cgs system. One form of energy that is well known is the kinetic energy and it's given by $1/2mv^2$. In cgs units, $m = g$, $v = cm\ sec^{-1}$ and $v^2 = cm^2 sec^{-2}$. Consequently, kinetic energy $=1/2mv^2 = g\ cm^2 sec^{-2}$ But $g\ cm^2\ sec^{-2}$ = erg. Therefore, the units cancel to give

$$T = \left[\frac{2.42 \times 10^{12}}{24.9 \times 10^7}\right] deg$$

This is absolute (as usual).

$$T = \left[\frac{2.42}{24.9} \times 10^5\right]^0 K = 0.0972 \times 10^5 °K$$

$$= 9.72 \times 10^3 °K$$

Solution 18. Assume a one dimensional problem; consequently only collisions in the positive x-direction will be considered. The mean velocity is given by the integral of $cx\, f(cx)$ between the limits of $c_x = 0$ and $c_x = \infty$ That is,

$$\bar{c}_x = \int_0^\infty c_x f(c_x)\, dc_x \qquad (1)$$

where $f(cx)$ is some exponetial functin of cx given by

$$f(cx) = d\left(\frac{m}{2\pi kT}\right)^{1/2} \exp\left(\frac{-\tfrac{1}{2} mc_x^2}{kT}\right) \qquad (2)$$

Multiply the right-hand side of (2) by 2 so that $\int_0^\infty f(c_x)\, dc_x = 1$ Equations (1) and (2) becomes

$$\bar{c}_x = 2(m/2\pi kT)^{1/2} \int_0^\infty c_x \exp\left(\frac{-mc_x^2}{2kT}\right) dc_x$$

$$= 2\left(\frac{m}{2\pi kT}\right)^{1/2} (kT/m)$$

$$= \sqrt{\frac{2kT}{\pi m}}$$

$$\bar{c}_x = \sqrt{\frac{2RT}{\pi M}}$$

The collision frequency, Z is given by

where N is the number of molecules and $\frac{1}{2}$ is the fraction of molecules moving toward the surface.

$$Z = \frac{\tfrac{1}{2} \times 1 \text{ atm} \times 6.023 \times 10^{23} \text{ molecules mole}^{-1}}{82.06 \text{ cm}^3 \text{ atm deg}^{-1} \text{ mole}^{-1} \times 298 \text{deg}}$$

$$\times \sqrt{\frac{2 \times 8.314 \times 10^7 \times 298}{\pi}} \left(\frac{0.2}{\sqrt{32}} + \frac{0.8}{\sqrt{28}}\right)$$

$= 2.885 \times 10^{23}$ collisions cm^{-2} sec^{-1}

But the number of collisions = ZA, where A is the area.

$$ZA = \left(2.885 \times 10^{23} \text{ collisions cm}^{-2} \text{ sec}^{-1}\right)\left(\frac{100 \text{ cm}}{1\text{m}}\right)^2 (2.0\text{m}^2)$$

or

$$ZA = 5.77 \times 10^{27} \text{ collisions sec}^{-1}$$

Solution 19. (*a*) The number of collisions a single atom makes per unit time is called the collision frequency and is given by

$$Z = \sqrt{2}\,\sigma \bar{C} N/V \tag{1}$$

where is the area of oven = d^2 and d is the cross-section. $\bar{C}$ is the mean speed of the atoms, V is the volume of the oven and N is the number of molecules per unit volume.

From equation (1)

$$\frac{N}{V} = \frac{nL}{V} = \frac{P}{kT}$$

where P is the pressure

$$Z = 2^{1/2}\ \bar{C}\left(\frac{P}{kT}\right)$$

$$\sigma = \pi d^2$$

The cross-section of the oven is given as 5.40 Å = $540 \times 10^{-12} m$

$$\sigma = \pi \times (540 \times 10^{-12} m)^2$$

$$\sigma = 9.2 \times 10^{-19} m^2$$

Also

$$P = \left(\frac{80\ mm\ Hg}{760\ mm\ Hg}\right)(1.0133 \times 10^5\ Nm^{-2)}$$

$$= 1.07 \times 10^4\ Nm^{-2}$$

Substituting for P, σ and $\bar{C}$ in equation (1) gives

$$Z = \frac{2^{1/2}\,(9.2 \times 10^{-19}\ m^{2)}\,(351\ m - s^{-1)}\,(1.07 \times 10^4\ Nm^{-2)}}{1.381 \times 10^{-23}\ JK^{-1)}\,(773°K)}$$

or

$$Z = 4.6 \times 10^8\ s^{-1}$$

(*b*) The total number of atomic collisions is given by

$$Z_{CsCs} = \tfrac{1}{2} Z\, N/V/\tfrac{1}{2} Z(P/kT)$$

$$= \frac{½(4.6 \times 10^{8}\, s^{-1})\,(1.07 \times 10^{4}\, Nm^{-2})}{1.38 \times 10^{-23}\, JK^{-1})\,(773°K)}$$

$$=(2.31 \times 10^{32} s^{-1}\, m^{-3}\,)\left(\frac{1m}{100\, cm}\right)^{3}$$

$$Z_{CsCs} = 1.16 \times 10^{28}\ \sec^{-1}$$

Solution 20. A gas in two dimensions comprise a system of molecules having kinetic energy due to motion along two directions in space, say the *x*-and *y*-directions. In this problems collisions are directed at a 1 cm length, therefore only components of velocity in the positive *x*-direction are considered.

Assume there are N molecules per unit area in the container then from the Boltzman equation, the number of molecules with velocities between *u* and *u*+*du* are.

$$NP(u)du.$$

The total number of molecules striking the unit area of the container per second in this velocity range are

$$N\,_{u}P(u)du \qquad (1)$$

Therefore the collision frequency, Z, per unit length is obtained by integrating (1) between the limits of $u = 0$ and $u = +\infty$ Thus,

$$Z = N\int_{0}^{\infty} uP(u)du$$

The integral can be evaluated easily using the explicit form of the velocity distributions.

$$Z = N\left(\frac{m}{2\,\pi\, kT}\right)^{½}\int_{0}^{\infty} \exp\left[\frac{-mu^{2}}{2kT}\right] u\, du$$

$$= N\left(\frac{m}{2\pi kT}\right)^{½}\left(\frac{2kT}{2\,m}\right)$$

$$Z = N\left(\frac{kT}{2\pi m}\right)^{½}$$

The two dimensional average speed follows in the same manner and is given by

$$\bar{c} = \int_0^\infty p\,(c)\,cdc$$

$$= \frac{m}{2k}\int_0^\infty c^2 \exp\left[\frac{-mc^2}{2kT}\right] dc$$

$$= \frac{m}{kT}\frac{\pi}{4}\left(\frac{2kT}{m}\right)^{3/2}$$

$$\bar{c} = \left(\frac{\pi kT}{2\,m}\right)^{1/2}$$

Therefore

$$\frac{\bar{c}}{\pi} = \left(\frac{kT}{2\,m}\right)^{1/2}$$

or $$Z = \frac{N\bar{c}}{\pi}$$

Solution 21. (*a*) The mean free path, λ, is a measure of the average distance that a molecule travels between successive collisions and is defined as

$$\lambda = \frac{\bar{c}}{Z} \qquad (1)$$

where $\bar{c}$ is the mean speed of the atoms and Z is the collision frequency.

Equation (1) can also be written as

$$\lambda = \frac{\bar{c}}{2^{1/2}\pi\sigma^2\bar{c}c}$$

or

$$\lambda = \frac{1}{2^{1/2}\pi\sigma^2\bar{c}c}$$

where c is the concentration

$$\lambda = \frac{1}{2^{1/2}\pi\,(10^{-5}\text{ atom }cm^{-3})\;(4\times10^{-16}\;cm^{2})}\left(\frac{1\text{ light–yr}}{9.464\times10^{17}\;cm}\right)$$

$$\lambda = 59\text{ light-yr}$$

(*b*) The mean speed is given by

$$\bar{c} = \sqrt{\frac{8RT}{\pi M}}$$

where M is the molecular weight of hydrogen and R is the gas constant. Inserting the respective values into the equation yields

$$\bar{c} = \sqrt{\frac{8(8.314)\,(10^7)\,(7.8 \times 10^5 \text{ erg } (g \text{ atom})^{-1}}{\pi\,(1.008g\,(g \text{ atom})^{-1}}}$$

$$= \sqrt{1.64 \times 10^{14} \text{ erg } g^{-1}}$$

$$\bar{c} = 1.28 \times 10^7 \text{ cm sec}^{-1}$$

$$\text{Time, } t = \frac{\lambda}{\chi}$$

$$= \frac{59 \text{ light–yr } (9.464 \times 10^{17} \text{ cm}(1 \text{ light–yr})^{-1})}{1.28 \times 10^7 \text{ cm sec}^{-1}} \left(\frac{1 \text{ yr}}{3.156 \times 10^7 \text{ sec}}\right)$$

$$= 138.2 \times 10^3 \text{ yr}$$

or

$$t = 1.382 \times 10^5 \text{ yr}$$

Solution 22. The V = f(P,T) expression means that molar volume is a function of pressure and temperature. If the molar volume is a property of the system, then equation (1) can be differentiated to give

$$dV = \left(\frac{\partial V}{\partial P}\right)_T dP + \left(\frac{\partial V}{\partial T}\right)_P dT \qquad (1)$$

Since V is a function of pressure and temperature and an ideal gas is being considered, the appropriate formulas to use are Boyle's and Gay-Lussac's laws. These are VP = k and V = k′T respectively. They will be useful in evaluating $(\partial/\partial P)_T$ and $(\partial/\partial T)_P$. Note that k and k^1 are proportionality constants. From Boyle's law, $(\partial/\partial P)_T = -k/P^2$, But k = VP. Therefore,

$$\left(\frac{\partial V}{\partial P}\right)_T = -\frac{VP}{P^2} = -\frac{V}{P} \qquad (2)$$

and from Gay-Lussac's law,

$$\left(\frac{\partial V}{\partial T}\right) = k' = \frac{V}{T}$$

Substituting these into equation (2) gives

$$dV = -\frac{V}{P}dP + \frac{V}{T}dT$$

and separating variables yields

$$\frac{dV}{V} = -\frac{dP}{P} + \frac{dT}{T}$$

or

$$\frac{dV}{V} + \frac{dP}{P} - \frac{dT}{T} = 0 \qquad (3)$$

Integrating equation (3) gives

$$ln V + ln P - ln T = ln R$$

where R = integration constant.

Equation (4) can also be put in the form

$$ln\left(\frac{VP}{T}\right) = ln R$$

Taking the exponents of both sides gives

$$e^{l ln(VP/T)} = e^{ln R}$$

or

$$\frac{VP}{T} = R\,.$$

Note that R is the universal gas constant.

Solution 23. Because the gas is ideal, the Ideal Gas law applies and the problem becomes merely an exercise in choosing the correct units. These units can be chosen correctly by converting the units of the values given to the units of the gas constant, R. Therefore, the volume must be in liters and the pressure, in atm. The temperature must be absolute (°K). The ideal gas law is represented by the formula $PV = nRT$ where P is the pressure in atmospheres, V is the volume in liters, n is the number of moles, R is the Universal Gas constant in liters-atm/mole-°K, and T is in degrees kelvin. Thus,

$$\frac{380 \text{ torr}}{760 \text{ torr/atm}} = .500 \text{ atm}, 27^\circ\text{C} + 273 = 300^\circ\text{K}$$

$$492 \text{ cm3} \times \frac{1 \text{ liter}}{1000 \text{ cm}^3} = .492 \text{ liters.}$$

$$\text{PV} = n\text{RT or n} = \frac{\text{PV}}{\text{RT}} = \frac{.500 \text{ atim} \times .492 \text{ liters}}{.0821 \text{ liters atm/mole}^\circ\text{K} \times 300^\circ\text{K}}$$

$$= .\text{moles.}$$

Solution 24. The column of air at a uniform temperature is subject to a gravitational field of acceleration, g. Due to this field, the forces acting upon the lower and upper boundaries are not the same. The difference is equal to the weight of the fluid between the boundaries.

Thus,

$$-df = \pi r^2 \text{ P}g\ dh \qquad (1)$$

where f = force, ρ = density, dh = difference in height and g = gravity constant. But the difference in the force per unit area is equal to the pressure difference between the lower and upper boundaries. Therefore, equation (1) takes the form

$$\frac{-df}{\pi r^2} = -dP = \rho g dh \qquad (2)$$

The minus sign indicates that the pressure decreases with increasing altitude. For ideal gases,

$$\rho = \frac{\text{PM}}{\text{RT}}$$

where M = molecular weight.

Using equation (3) appropriately, equation (2) becomes

$$-dP = \frac{\text{PM}}{\text{RT}}\ gdh$$

$$\frac{dP}{\text{P}} = -\frac{\text{M}g}{\text{RT}}\ dh \qquad (4)$$

Integrating equation (4) gives

$$\int_{P_1}^{P_2} \frac{dP}{P} = -\frac{Mg}{RT}\int_0^h dh$$

But $P_1 = 1$ atmosphere. Therefore

$$\int_1^{P_2} \frac{dP}{P} = -\frac{Mg}{RT} h$$

or

$$\ln P \Big|_1^{P_2} = -\frac{Mg}{RT} h ,$$

which becomes

$$\ln\left(\frac{P_2}{1}\right) = -\frac{Mg}{RT} h$$

$$= -\frac{(29)(980)(1.5\times 10^6)}{(8.31\times 10^7)(273)}$$

$$= -1.879$$

$$P_2 = e^{-1.879} = 0.153 \text{ atm}$$

Since the weight of a column of air extending down to sea level is 1×10^6 dynes cm^{-2} , the weight of a column extending down only to a given altitude will be 0.153×10^6 dynes cm^{-2}. Therefore, the weight of the column between the sea level and this given altitude is $1 \times 10^6 - 0.153 \times 10^6 = 0.847 \times 10^6$ dynes cm^{-2}. The weight of this column between the sea level and the given altitude is given by

$$w = nMg \tag{5}$$

where w = weight

n = number of moles.

Rearranging equation (5) to solve for n gives

$$n = \frac{w}{Mg} = \frac{(0.847\times 10^6)}{(29)(980)}$$

$$n = 29.8 \text{ moles.}$$

Solution 25. Using the ideal gas formula, we have

$$PV = nRT$$

and $$n = \frac{PV}{RT} \tag{2}$$

After the addition of oxygen equation (2) becomes

$$(n + .01) = \frac{PV}{RT} = \frac{(0.5)\,(2)}{(0.082)\,(283)} = 0.432$$

$$n = 0.0432 - 0.01 = 0.0332$$

Now, use equation (1) PV = nRT, to solve for T

$$(0.5)\,(2) = (0.0332)\,(0.082)\ T$$

$$T = \frac{(.5)\,(2)}{(0.0332)\,(0.082)} = 367^{\circ}K$$

Solution 26. In taking a first look at the problem it might seem out of place because of the word concentration. However, its units are moles/liter. In the ideal gas law n = number of moles and liters represent the volume. Therefore,

$$c = 1 \text{ mole liter}^{-1} = \frac{n}{v}$$

Where c = concentration. The ideal gas law is defined as

$$PV = nRT \tag{1}$$

where P = Pressure, V = volume, R = gas constant and T = temperature. Rearranging equation (1) to solve for P gives

$$P = \frac{n}{v}\ RT. \tag{2}$$

Observe that the n/V term in equation (2) is the definition of concentration. Therefore,

$$P = cRT$$

or $$1 \text{ atm} = 1\ \frac{\text{mole}}{\text{liter}}\ RT. \tag{3}$$

The R in the right side of equation (2) has units of 0.082

$\frac{\text{liter atm}}{\text{mole deg}^\circ\text{K}}$

Hence equation (3) becomes;

$$1 \text{ atm} = \left(1\frac{\text{mole}}{\text{liter}}\right)\left(0.082\ \frac{\text{liter atm}}{\text{mole deg}}\right)T$$

Solving for T gives

$$T = \left[\frac{1}{0.082}\right] \text{deg.}$$

Note that deg is always absolute

$$T = 12.19^\circ\text{K}$$

Consequently, an ideal gas will exhibit a pressure of 1 atm and a concentration of mole liter^{-1} only when the absolute temperature is 12.19°K.

Solution 27 (*a*) Consider a column of air of unit cross-sectional area, and *dx* cm thick at an altitude, *x*, above the surface of the earth. Due to gravitational forces, the density and consequently the pressure of the earth's atmosphere is not uniform. The pressure decreases with increasing altitude. Therefore, the forces acting upon the top and bottom boundaries of the disc are not the same. This difference in these forces is equal to the weight of the disc. Thus,

$$-df = \rho g dx$$

The difference in the force per unit area is equal to the pressure difference between the top and bottom boundaries of the disc:

$$df = -dp = \rho\ gdx \qquad (1)$$

where g is the acceleration due of gravity and p is density. Notice the minus sign which indicates that the pressure decreases with increasing altitude. Equation (1) may be rewritten as

$$df = dp = -\ \rho g dx. \qquad (2)$$

But $\rho = PM/RT$

where M is the average molecular mass of air and R is the gas constant, therefore equation (2) changes to

$$dp = \frac{-PM}{RT}\, gdx. \qquad (2)$$

Applying the boundary condition that at $x = 0$, $P = P_0$. Equation (2) can be integrated after separating the variables and using the appropriate limits. Thus,

$$\int_{P_0}^{P} \frac{dP}{P} = -\frac{Mg}{RT}\int_0^x dx$$

$$ln\ P\ |_{P_0}^{P} = -\frac{Mg}{RT}\,x$$

or $$ln\ \frac{P}{P_0} = -\frac{Mgx}{RT}$$

Taking the exponent of both sides gives

$$\frac{P}{P_0} = e^{-Mgx/RT}$$

or $$P = P_0\, e^{-Mgx/RT} \qquad (4)$$

(*b*) Using $x = 8km$, $P_0 = 760mm$ Hg and $T = 273°K$ in equation (4) gives

$$P = 760 \text{ mm Hg } e^{\frac{-\left(29\text{gmole}^{-1}\right)\left(\frac{1\text{kg}}{1000\text{g}}\right)\left(9.8\text{m sec}^{-2}\right)(8000\text{m})}{\left(8.314\text{J}°\text{K}^{-1}\text{ mole}^{-1}\right)(273°\text{K})}}$$

$$= 760 \text{ mm Hg}$$

$$\exp \frac{-\left(29\text{gmole}^{-1}\right)\left(\frac{1\text{kg}}{1000\text{g}}\right)\left(9.8\text{m sec}^{-2}\right)(8000\text{m})}{\left(8.314\frac{\text{m}^2\text{kg}}{\text{sec}^2}°K^{-1}\text{mole}^{-1}\right)(271°\text{K})}$$

The exponential term is dimensionless as expected. Thus,

$$P = 760 \text{ mm Hg} \exp(-1.0017)$$

or $$= 760 \text{ mm Hg } (0.3672)$$

$$P = 279 \text{ mm Hg}.$$

Solution 28. The ideal gas law is given by $PV = nRT$. In this

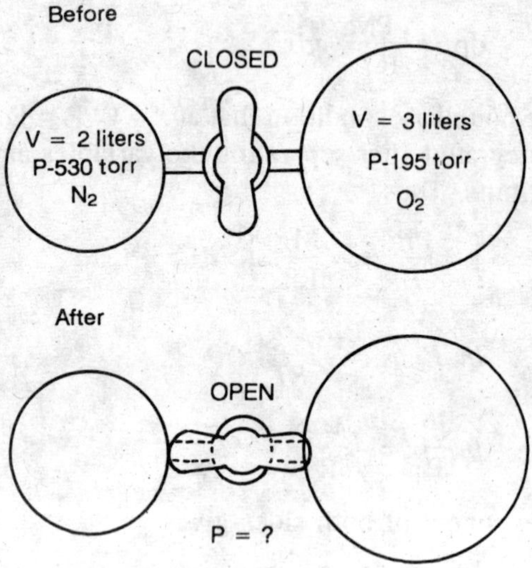

Fig.

problem,

P = final pressure

V = final volume

n = total moles of oxygen and nitrogen

R = universal gas constant

T = temperature.

By Dalton's law of partial pressures, the total final pressure is equal to final pressure of oxygen plus the final pressure of nitrogen.

Therefore,

$$P_{\text{final}} \text{ js } P_{O_2} + P_{N_2} = n_{O_2}\frac{RT}{V} + n_{N_2}\frac{RT}{v} \qquad (1)$$

Note that V = 3 + 2 = 5 liters because the P_{final} of the system is being considered after equilibriation.

Note that T is not known. It is assumed to be constant and may cancel.

Using original conditions, n_{O_2} and n_{N_2} can be calculated as follows:

and $$n_{O_2} = \frac{P_{O_2}V_{O_2}}{RT} = \frac{(195\text{ torr})\ (3\text{ liters})}{RT}$$

$$n_{o_2} = \frac{P_{N_2}V_{N_2}}{RT} = \frac{(530\text{ torr})\ (2\text{ liters})}{RT}$$

From these, $n_{O_2}RT = P_{O_2}V_{O_2}$ and $n_{N_2}RT = P_{N_2}V_{N_2}$. Adding up both equations give

$$n_{N_2}RT + n_{N_2}RT = P_{O_2}V_{O_2} + P_{N_2}V_{N_2}$$

or $$RT\,(n_{O_2} + n_{N_2}) = (195)\ (3) + (530)\ (2)$$

Observe that if this equation is divided by V, it becomes equation (1) which is the equation for solving for P_{final}. That is,

$$\frac{RT}{V}(n_{O_2} + n_{N_2}) = \frac{(195)\ (3) + (530)\ (2)}{V}$$

Recall that V = 5. Therefore,

$$P_{final} = \frac{585 + 1060}{5} = 329\text{ torr.}$$

Solution 29. In this problem, the number of moles of gas remains the same. Therefore, using n = PV/RT, when the two flasks are immersed in boiling water,

$$n = n_1 + n_2 = \frac{P}{R}\frac{V_1}{T_1} + \frac{P}{R}\frac{V_2}{T_2} + \frac{P}{R}\left(\frac{V_1}{T_1} + \frac{V_2}{T_2}\right)$$

But $V_1 = V_2 = V$ at the initial stage

$$n = \frac{.5}{R}\left(\frac{V}{373} + \frac{V}{373}\right) = \frac{.5}{R}\left(\frac{2V}{373}\right) = \frac{.5V}{R}\left(\frac{2}{373}\right)$$

when one of the flasks is immersed in an ice-water mixture

$$n =$$

$$n_1' + n_2' = \frac{P'}{R}\left(\frac{V}{273} + \frac{V}{373}\right) = \frac{P'V}{R}\left(\frac{1}{273} + \frac{1}{373}\right)$$

Now n_1' = number of moles of gas in flask 1 immersed in ice-water mixture

n_2 = number of moles of gas in flask 2

Equating both mathematical expressions for n,

$$\frac{.5V}{R}\left(\frac{2}{373}\right) = \frac{P'V}{R}\left(\frac{1}{273} + \frac{1}{373}\right)$$

$$P' = \left[\frac{.5V}{R}\left(\frac{2}{373}\right)\right] \Big/ \left[\frac{V}{R}\left(\frac{1}{273} + \frac{1}{373}\right)\right]$$

$$= [(0.5)\ (546)]\ /\ (646) = 0.423 \text{ atm.}$$

Solution 30. The volume of helium flowing through the wall of the flask varies linearly with the partial pressure difference and inversely with the thickness of the wall. Define dV_s/dt = rate of flow in cc (STP) per sec.

$$\frac{dV_s}{dt} = \frac{k(P_0 - P)A}{d} \qquad (1)$$

where k = permeability (*cc* gas (STP) per sec per cm^2 of surface per millimeter thickness per millibar partial pressure difference)

P_0 = pressure of helium outside of flask (m bar)

P_1 = pressure of helium inside of flask (m bar)

A = surface area of flask (cm2)

d = thickness of wall (mm)

$$V_s = \frac{VT_sP}{P_sT}$$

$$V_s = \frac{(100 \text{ cc})\ (3\backslash 273.15^\circ\text{K})\ (\text{P m bar})}{(1013.3 \text{ m bar atm}^{-1})\ (1 \text{ atm})\ (293.15^\circ\text{K})}$$

$$V_s = 0.09125 \text{ P}$$

Taking derivatives of both sides with respect to time yields,

$$\frac{dVs}{dt} = 0.0915 \text{ dP}/dt \qquad (2)$$

Combining equations (1) and (2) gives,

$$\frac{dP}{dt} = \frac{k(P_0 - P)A}{0.0915d} \tag{3}$$

By definition,

$$A = 4\pi r^2 \text{ and } V = 4/3\ \pi r^3$$

Solving for r from the V expression gives

$$r^3 = \frac{3V}{4\pi}$$

or
$$r = \left(\frac{3V}{3\pi}\right)^{1/3}$$

Squaring r yields

$$r^2 = \left(\frac{3V}{4\pi}\right)^{2/3}$$

Substituting this into the A expression yields

$$A = 4\pi\left(\frac{3V}{4\pi}\right)^{2/3} \quad 4\pi\left(\frac{3\times 100}{4\pi}\right)^{2/3}$$

$$A = 104.2\ \text{cm}^2$$

From equation (3)

$$\frac{dP}{(P_0 - p)} = \frac{kA}{0.0915d}\, dt$$

$$\frac{dp}{(P_0 - P)} = \frac{(4.9\times 10^{-13})\,(104.2\text{cm}^2)}{(0.0915)\,(0.7)}\, dt \tag{4}$$

Intergrating equation (4) gives,

$$\int_0^{P_0} \frac{dP}{P_0 - P} = 7.932\times 10^{-10} \int_0^t dt$$

$$l\frac{P_0}{P_0 - P} = 7.932\times 10^{-10}\, t$$

$$t = 30\times 24\times 3600$$

$$= 2.592\times 10^6 \text{ sec}$$

$$P_0 = X_{He}\, P_{total}$$
$$= (5 \times 10^{-6})\,(1013.3)$$
$$= 5.066 \times 10^{-3} \text{ m bar}$$

$$\frac{P_0}{P_0 - P} = e\,(7.932 \times 10^{-10})\,(2.592 \times 10^{-6})$$
$$= 1.002058$$

$$P_0 = 1.002058\,(P_0 - P)$$
$$P_0 = 1.002058\, P_0 - 1.002058\, P$$
$$1.002058\, P = P_0\,(1.002058 - 1)$$
$$1.002058\, P = 5.066 \times 10^{-3} \text{ m bar } (0.002058)$$
$$1.002058\, P = 1.0426 \times 10^{-5}$$
$$P = 1.04 \times 10^{-5} \text{ m bar}$$

Solution 31. From one season to another, the only properties of the gas that will change are pressure and temperature. The mass (hence the number of moles) and the volume will remain the same. If it is assumed that this gas is ideal, then

$$PV = nRT \tag{1}$$

where
P = Pressure of the gas
V = Volume of the gas
n = number of moles
R = gas constant
T = Temperature of the gas.

Rearranging equation (1) to solve for P gives,

$$P = (n/V)\, RT. \tag{2}$$

Since n and V are constant, equation (2) shows that pressure is directly proportional to temperature. That is, P/T = nR/V = constant. Therefore

$$\frac{P_1}{T_1} = \frac{P_2}{T_2} = \frac{n_1 R}{V_1} = \frac{n_2 R}{V_2} \tag{3}$$

where P_1 = initial pressure

T_1 = initial temperature

P_2 = final pressure

T_2 = final temperature

n_1 and n_2 = intial and final moles respectively. V_1 and V_2 = initial and final volume respectively.

The moles and volume are not changing; therefore, $n_1 = n_2$ and $V_1 = V_2$. Consequently, equation (3) can be written as

$$\frac{P_1}{T_1} = \frac{P_2}{T_2} \quad (4)$$

Before equation (4) can be used, the pressure and temperature must be in absolute scales.

$$\frac{{}^\circ C}{5} = \frac{{}^\circ F - 32}{9} \quad (5)$$

and $$P = 14.7 \text{ psia} + \text{psig} \quad (6)$$

where °C = degrees centigrade

°F = degrees farenheit

Psia = absolute psi

Psig = gauge psi

Using equation (5) and (6)

$$122^\circ F = 50^\circ C = (50 + 273)^\circ K = 323^\circ K$$

and

$$P = 14.7 + 30 = 44.7 \text{ psia.}$$

These can now be inserted into equation (4) to give,

$$\frac{44.7}{273} = \frac{P_2}{323}$$

Therefore,

$$P_2 = \left[\frac{(44.7)\,(323)}{273}\right] \text{psia}$$

$$= 52.9 \text{ psia}$$

or from equation (6),

$$52.9 \text{ psia} = 14.7 \text{ psia} + x \text{ psig}$$

$$P_2 = (52.9 - 14.7) \text{ psig}$$

$$= 38.2 \text{ psig}$$

Solution 32. The procedure used is to compare the two cases that involve the same volume and pressure. Therefore,

$$n_1T_1 = n_2T_2 \qquad (1)$$

where

n_1 = moles of first gas

n_2 = moles of second gas

The molecular weight of chlorine is 71. Therefore 7.1 gm corresponds to

$$(7.1 \text{ g})\left(\frac{\text{mole}}{71g}\right) = 0.1 \text{ mole.}$$

Since the final 6.4 gm corresponds to 10% less,

$$n2 = .1 - 10\% \text{ of } .1$$

$$= .1 - (10/100)(.1) = .1 - .01$$

$$= .09 \text{ moles.}$$

Using equation (1), $.1T_1 = .09T_2$. But $T_2 = T_1 + 30$

$$.1T_1 = .09(T_1 + 30)$$

$$.1T_1 = .09T_1 + 2.7$$

$$.T_1 - .09T_1 = 2.7$$

$$0.01T_1 = 2.7$$

$$T_1 = 2.7/.01 = 270^\circ\text{K}$$

Now V = 2.24 liters. Using PV = nRT, it follows that

$$P = \frac{nRT}{V}$$

$$= \frac{(0.1\text{mole})\left(\frac{0.082\ \text{l–atm}}{^\circ\text{Kmole}}\right)(270^\circ\text{K})}{22.4\text{l}}$$

$$P = 0.99\ \text{atm}$$

Solution 33. Using PV = nRT, $P_AV_A = n_ART_A$ for gas A and $P_BV_B = n_BRT_B$ for gas B.

The problem states that $T_A = T_B = T$. So $P_AV_A = n_ART$ and $P_BV_B = n_BRT$ after substitution. Values for n and V are not given. However, it is known that

density = mass/volume

volume = mass/density

Now $$P_A\left(\frac{M_A}{P_A}\right) = \left(\frac{M_A}{MW_A}\right)RT$$

and

$$P_B\left(\frac{M_B}{P_B}\right) = \left(\frac{M_B}{MW_B}\right)RT$$

where M = Mass

MW = Molecular Weigh

ρ = Density

The problem also states that $\rho_A = 2\rho B$ thus, $\rho B = p_A/^2$, $MW_A = ½\ MW_B$ and $MW_B = 2MW_A$

Using the ideal equation with appropriate substitution,

$$P_A\left(\frac{M_A}{\rho_A}\right) = \left(\frac{M_A}{MW_A}\right)RT \qquad (1)$$

and $$P_B\left(\frac{M_B}{\rho A/2}\right) = \left(\frac{M_B}{2MW_A}\right)RT \qquad (2)$$

Dividing (1) by (2),

$$\frac{P_A\left(\frac{M_A}{\rho_A}\right)}{P_B\left(\frac{2M_B}{\rho_A}\right)} = \frac{\left(\frac{M_A}{MW_A}\right)RT}{\left(\frac{M_B}{2MW_A}\right)RT}$$

$$\frac{P_A}{P_B} = \left[\left(\frac{2M_B}{\rho_{A)}}\right)\left(\frac{M_A}{MW_A}\right)\right] \Big/ \left[\left(\frac{M_A}{\rho_A}\right)\left(\frac{M_B}{2MW_A}\right)\right] = \left[\frac{2M_BM_A}{\rho_A MW_A}\right]$$

$$\Big/ \left[\frac{M_AM_B}{\rho_A 2MW_A}\right]$$

$$= \frac{2M_BM_A}{P_AMW_A} \times \frac{\rho_A 2MW_A}{M_AM_B} = 4$$

$$P_A/P_B = 4$$

Solution 34. The equation $P_1V_1/T_1 = P_2V_2/T_2$ is applicable here. But $T_1 = T_2$, $P_1T_1 = P_2V_2$

where P_1 = initial pressure

P_2 = final pressure

V_1 = initial volume

V_2 = final volume

$$P_2 = P_1\left(\frac{V_1}{V_2}\right) = P_1\left(\frac{2 \times \text{area of pipe}}{h \times \text{area of pipe}}\right)$$

$$P_2 = 2P_1/h.$$

In this problem P_1 = 1 atm = 10m hydrostatic head and P_2 = (10 + h) metres. From equation (1) we have

$$10 + h\ \frac{2(10)}{h} = \frac{20}{h}$$

$$h\ (10 + h) = 20$$

$$h^2 + 10h = 20$$

$$h^2 + 10h - 20 = 0$$

This is a quadratic equation and there are two possible values

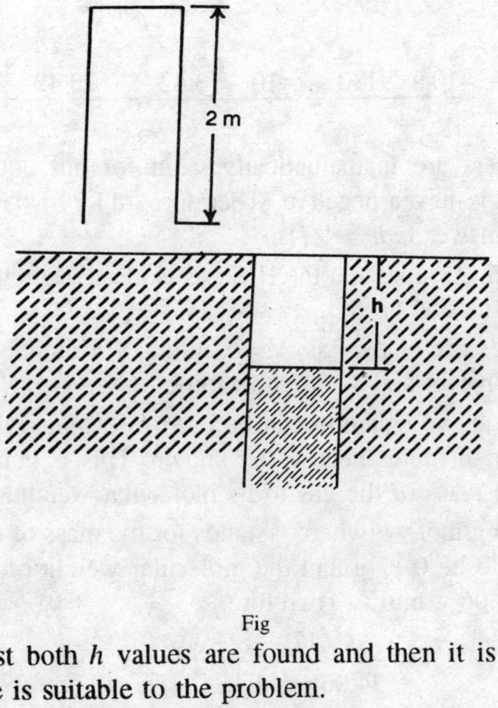

Fig

for h. First both h values are found and then it is determined which one is suitable to the problem.

Using the "almighty" formula for solving quardratic equations,

$$h = \frac{-b \pm \sqrt{b^2 - 4ac}}{2a}$$

where b = coefficient of h = 10 in this case

a = coefficient of h^2 = 1 in this case

c = the constant = –20 int this case.

$$h = \frac{-10 \pm \sqrt{10^2 - (4)(1)(-20)}}{2(1)}$$

$$= \frac{-10 \pm \sqrt{100} + 80}{2} = \frac{-10 \pm \sqrt{180}}{2}$$

The answers are either

$$= \frac{10 + \sqrt{180}}{2} = \frac{3.42}{2} = 1.71\text{m},$$

or

$$\frac{-10 - \sqrt{180}}{2} = \frac{-10 - 13.42}{2} = \frac{-23.42}{2} = -11.71\text{m.t}$$

Both answers are mathematically valid for our equation. But physically, is never negative. Therefore −11.71m is discarded. The final answer is h = 1.71m.

Solution 35. Assume that the H_2S gas is ideal. Then the relation, PV = nRT can be used.

P is the pressure of the gas, V = volume, n = number of moles, R = gas constant and T = temperature.

Before the above equation can be used to evaluate the volume, the number of moles need to be known. This is defined as the ratio of the mass of the gas to its molecular weight.

That is, n = g/mol.wt. where g stands for the mass of the sample. g is given to be 0.17 g and the molecular weight of H_2S = 2 + 32.06 = 34.06 g mol^{-1} Therefore,

$$n = \frac{0.17g}{34.06g\ \text{mol}^{-1}}$$

$$= 0.005\ \text{mol}$$

Solving the ideal gas law for V gives

$$V = \frac{nRT}{P} = \frac{(0.005)RT}{P} \qquad (1)$$

R = 0.082 liter atm $\text{mol}^{-1}\ \text{k}^{-1}$. Since R is in liter atm $\text{mol}^{-1}\ \text{K}^{-1}$, T has to be converted to K and P to atm. Thus,

$$T = 27^\circ\text{C} = (27 + 273)^\circ\text{K} = 300^\circ\text{K}$$

and

$$P = 380\ \text{torr} = \frac{380\ \text{torr}}{760\ \text{torr atm}^{-1}} = 0.50\ \text{atm}.$$

Substituting these into equation (1) gives

$$V = \frac{(0.005 \text{ mol})\ (0.082 \text{ liter atm mol}^{-1}\ \text{K}^{-1})\ (300\text{K})}{(0.50 \text{ atm})}$$

Observe that all the units cancel out except the liter, which is what the problem requires.

$$V = 0.246 \text{ liter.}$$

Solution 36. Here, the properties of the oil that are changing are volume and temperature. If this is ideal, then

$$PV = nRT \tag{1}$$

where P = pressure, V = volume, n = number of moles, R = constant and T = temperature. Since n and P are constant,

$$V \propto T$$

or

$$\frac{V}{T} = \frac{nR}{P} \tag{2}$$

Equation (2) can be written in the form

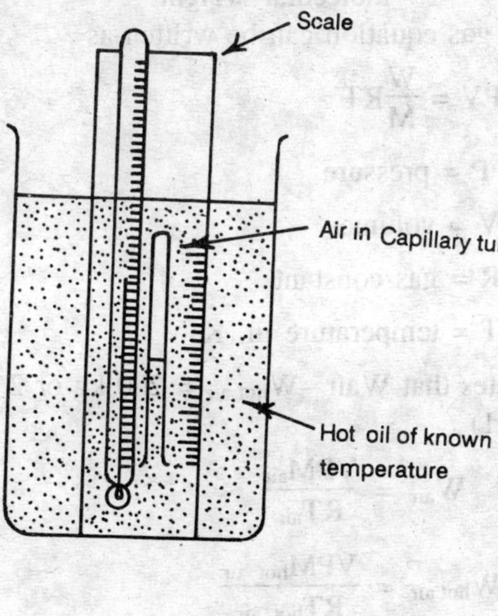

Fig.

$$\frac{V_1}{T_1} = \frac{V_2}{T_2} = \frac{n_1R}{P_1} = \frac{n_2R}{P_2} \quad (3)$$

But $n_1 = n_2$ and $P_1 = P_2$. Therefore equation (3) becomes

$$\frac{V_1}{T_1} = \frac{V_2}{T_2} \quad (4)$$

where V_1 and V_2 = initial and final volumes respectively and T_1 and T_2 = initial and final temperature respectively. Since T is supposed to be in absolute scale, $T_1 = 68 + 273 = 341^\circ K$

$$T_2 = 136 + 273 = 409^\circ K$$

Using these in equation (4) yields

$$\frac{V_1}{V_2} = \frac{T_1}{T_2} = \frac{341}{409} = 0.833$$

$$V_2 = \frac{V_1}{0.833} = 1.2\ V_1$$

Thus the final volume is about 20% larger than the initial volume.

Solution 37. Since $n = \frac{\text{weight or mass}}{\text{molecular weight}} = W/M$ = number of moles, the ideal gas equation can be written as

$$PV = \frac{W}{M}RT$$

where P = pressure

V = volume

R = gas constant

T = temperature in °K

The problem states that $W_{air} - W_{hot\ air} = 200$ kg or 2×10^5 gm. From equation (1)

$$W_{air} = \frac{VPM_{air}}{RT_{air}}$$

$$W_{hot\ air} = \frac{VPM_{hot\ air}}{RT_{hot\ air}}$$

$$W_{air} - W_{hot\ air} = \frac{VP}{R}\left(\frac{M_{air}}{T_{air}} - \frac{M_{hot\ air}}{T_{hot\ air}}\right) = 2\times 10^5 gm$$

$$2\times 10^5 gm = \frac{VP}{R}\left(\frac{M_{air}}{T_{air}} - \frac{M_{hot\ air}}{T_{hot\ air}}\right)$$

and $$V = \frac{2\times 10^5 gm \times R}{P\left(\frac{M_{air}}{T_{air}} - \frac{M_{hot\ air}}{T_{hot\ air}}\right)}$$

$$= \frac{(2\times {}^5gm)\ (0\text{–}082 \text{ liters atm } {}^\circ K^{-1} \text{ mole}^{-1})}{1 \text{ atm}\left(\frac{29g \text{ mole}^{-1}}{298^\circ K} - \frac{32g \text{ mole}^{-1}}{373^\circ K}\right)}$$

$$= \frac{(2\times 10^5 gm)\ (0.082 \text{ liters atm}^\circ K^{-1} \text{ mole}^{-1})}{1 \text{ atm}(0.0115) \text{ g mole}^{-1}/^\circ K}$$

$$V = 1.42 \times 10^6 \text{ liters.}$$

Solution 38. There are 22.4 liters of air at STP (0°C and 1 atm). Therefore, there is 1 mole of air. Also, there are 4 gm of hydrogen, which is equivalent to

$$\frac{4gm}{2^{gm}/_{mole}} = 2 \text{ moles.}$$

By definition,

$$M_{average} = \left[\sum_{i=1}^{3} n_i M_i\right] \Big/ \left[\sum_{i=1}^{3} n_1\right]$$

where n_1 = moles of oxygen in air

n_2 = moles of nitrogen in air

n_3 = moles of hydrogen

M_1 = molecular weight of oxygen

M_2 = molecular weight of nitrogen

M_3 = molecular weight of hydrogen.

$$M_{average} = \frac{n_1M_1 + n_2M_2 + n_3M_3}{n_1 + n_2 + n_3}$$

$$= \frac{(.2)(32) + (.8)(28) + 2(2)}{.2 + .8 + 2} = \frac{32.8}{3} \text{ gm/mole}$$

$$= 10.9 \text{ gm/mole.}$$

Solution 39. The gas density = 0.002 gm/cm^3 = 0.002 gm/cm^3 $\times \dfrac{1000 \text{ cm}^3}{1 \text{ liter}}$ = 2gm/liter.

Note that this is nothing more than conversion of scales. Also 11 liters could have been converted to cm^3.

Since the gas density is 2 gm/liter (that is, 2 gm in every liter), the 11 liters will contain 22 gm; but the flask contains 20 gm of neon. Therefore, there are 22-20 = 2 gm of hydrogen. Therefore, there exists 1 mole of each gas or 2 moles totoal.

Thus, the average molecular weight = $\dfrac{20 + 2}{2}$ = 22/2 = 11gm/mole.

Using PV = nRT which upon rearranging and substitution yields.

$$P = \frac{(2)(0.082)(273)}{11} = 4.07 \text{ atm.}$$

Remember that temperature is in absolute kelvin.

Solution 40. Define W_{Ar} and W_{He} as the weight of argon and helium in grams respectively. Define also P_{Ar} and P_{He} as the partial pressures of argon and helium respectively.

Therefore

$$W_{Ar} + W_{He} = 5.00 \text{ g} \qquad (1)$$

and

$$P_{Ar} + P_{He} = P_{total} = 1.00 \text{ atm} \qquad (2)$$

The molecular weight of a gas is given by

$$M = \frac{W}{n}$$

where W and n are the totoal weight and number of the moles respectively. From equation (2),

$$P = P_{Ar} + P_{He}$$

But P = nRT/V. Therefore,

$$P = n_{Ar}\frac{RT}{V} + n_{He}\frac{RT}{V}$$

From equation (3), n = W/M. Therefore, equation (4) becomes

$$P = \frac{W_{Ar}RT}{M_{Ar}V} + \frac{W_{He}RT}{M_{He}V}$$

From (1) $W_{Ar} = 5.00g - W_{He}$

$$\frac{\left(5.00g - W_{He}\right)RT}{\left(39.9g\ mole^{-1}\right)V} + \frac{W_{He}RT}{\left(4.00g\ mole^{-1}\right)V}$$

$$P = \left(\frac{5.00g - W_{He}}{39.9g\ mole^{-1}} + \frac{W_{He}}{400g\ mole^{-1}}\right)$$

$$\times \frac{\left(0.08211\ atm\ k^{-1}\ mol^{-1}\right)(298^{\circ}K)}{10.01}$$

$$P = 4.00\left[\frac{(5.00 - W_{He}) + 39.9W_{He}}{159.6}\right](2.44)$$

$$p = 1 atm$$

$$159.6 = (29 + 35\ W_{He})\ 2.44$$

$$159.6 = 48.8 + 87.5\ W_{He}$$

$$W_{He} = 1.26\ g$$

Weight per cent He $\frac{1.26g}{5.00g} \times 100 = 25.2\%$

Weight per cent Ar = 100 – 25.2

$= 74.8\ \%$

Solution 41. Use as a basis 1 mol hydrogen before dissociation (which has a weight of 2 gms). After 0.33 moles have dissociated, then we have

0.67 mole H_2 left.

Therefore, there are 0.66 moles of H. The total number of moles are 0.67 + 0.66 = 1.33 moles.
For a mixture of gases, the density

$$\rho = \frac{PM_{average}}{RT} \text{ where } M_{average} = \frac{\text{total weight}}{\text{total moles}} = \frac{2\text{gm}}{1.33 \text{ mol}}$$

Note that the units of $M_{average}$ are consistent with the units of molecular weight. This is very important.
Therefore,

$$\rho = \frac{(1)\ (2/1.33)}{(0.0082)\ (2273)} = 8.07 \times 10^{-3} \text{ gm/liter}$$

Solution 42. (a) The work done in an isothermal compression of an ideal gas is given by

$$W = -\int PdV \qquad (1)$$

But $P = \frac{nRT}{V}$. Substituting $P = n\ RT/V$ into equation (1) gives

$$W = -\int nRT\frac{dV}{V}$$

Since nRT is constant,

$$W = -nRT\int_{V_1}^{V_2} \frac{dV}{V}$$

or

$$W = -nRT \ln \frac{V_2}{V_1} \quad (2)$$

The number of moles, of C_2H_4 are

$$n = \left(1\text{kg } C_2H_4\right)\left(\frac{1000\text{g}}{1\text{kg}}\right)\left(\frac{1\text{mole } C_2H_4}{28.05\text{g}}\right)$$

$$n = 35.65 \text{ moles.}$$

Hence equation (2) becomes

$$W = -(35.65 \text{ moles})\left(8.314\text{J}^\circ\text{K}^{-1} \text{ mol}^{-1}\right)(300^\circ\text{K}) \ 1n \ \frac{10^{-2}\text{m}^3}{10^{-1}\text{m}^3}$$

$$= 2.047 \times 10^5 \text{ J}$$

or $$W = 204.7 \text{ kJ.}$$

(*b*) Again the work done is given by

$$W = -\int PdV$$

For a gas that obeys the Van der Waals equation, P is given by

$$P = \frac{nRT}{V - nb} - \frac{n^2a}{V^2}$$

where *a* and *b* are Van der Waals constants given as 4.47 l^2 atm mol^{-2} and 0.0571 mol^{-1} respectively.

Substituting for P gives

$$W = -\int PdV = -\int \left[\frac{nRT}{V - nb} - \frac{n^2a}{V^2}\right]dV$$

or $$W = -\int_{V_1}^{V_2} \frac{nRT}{V-nb}\, dV + \int_{V_2}^{V_1} \frac{n^2a}{V^2}\, dV$$

Taking out the constants from the integral gives

$$W = -n\, RT \int_{V_1}^{V_2} \frac{dV}{V-nb} + n^2a \int_{V_1}^{V_2} d\frac{V}{V^2} \quad (3)$$

To evaluate the integral

$$\int_1^2 \frac{dV}{V - nb},$$

let $y = V - nb$ where n and b are constants

$dy = dV$

$$\int_{y_1}^{y_2} \frac{dy}{y} = \ln y \Big|_{y_1}^{y_2} = \ln \frac{y_2}{y_1}$$

Equation (3) becomes

$$W = -nRT \ln \frac{V_{2-nb}}{V_{1-nb}} + n^2 a \left(\frac{V_2 - V_1}{V_2 V_1}\right)$$

$$= -(36.65 \text{ moles}) \left(8.314 \text{ JK}^{-1} \text{ mol}^{-1}\right) (300^\circ\text{K}) \ln \left(\frac{10^{-2} - (35.65)(0.0571 l \text{ mol}^{-1}}{10^{-1} - (35.65)(0.0571 l \text{ mol}^{-1}}\right)$$

$$+ (35.65 \text{ moles})^2 \left(4.47 l^2 \text{ atm mol}^{-2}\right) \left(\frac{10^{-2} - 10^{-1}}{(10^{-2})(10^{-1})}\right)$$

$$W = -515, 320 \ l\text{–atm}$$

Solution 43. If van der Waals forces are included the equation used is

$$\left(P + \frac{n^2 a}{V^2}\right)(V - nb) = nRT$$

where P = pressure

n = moles

V = volume

a and b = constants

R = gas constant

T = temperature

Looking at the above equation it is possible to compute (*a*) P, V, or T if the other two are known. (*b*) Its critical point. At the critical point the P, T and V in the equation are replaced by P_c, T_c and V_c, respectively. P_c is the critical pressure, V_c, the critical volume, and Tc, the critical temperature. Since the van der Waals constants, *a* and *b* are known it is possible to express the critical pressure, temperature and volume in terms of these constants. Thus

$$V_c = 3b; \quad P_c = \frac{a}{27b^2}\ ; \ T_c = \frac{8a}{27bR}$$

(*c*) The vapor pressure of the liquid; (*d*) Compressibility factor, Z. Z is written as $\frac{P_c V_c}{RT_c}$; (*e*) The heat of vaporization; (*f*) coefficient of thermal expansion.

Solution 44. If temperature, T and pressure, P are chosen to be independent variables, then, volume, V is some function of T and P. Thus,

$$V = f\,(T, P)$$

Writing the change in volume in differential form gives

$$dV = \left(\frac{\partial V}{\partial T}\right)_P dT + \left(\frac{\partial V}{\partial P}\right)_T dP \qquad (1)$$

The quantities $\left(\frac{\partial V}{\partial P}\right)_P$ and $\left(\frac{\partial V}{\partial P}\right)_T$ can be determined using α and β

$$\alpha = \frac{1}{V}\left(\frac{\partial V}{\partial T}\right) = k_1\left(\frac{C_P}{C_V}\right) T^{\,(C_P/C_V)-1}$$

or

$$\left(\frac{\partial V}{\partial T}\right)_P = \frac{K_1 C_P T^{\,(C_P/C_V)-1}\, V}{C_V} \qquad (2)$$

$$\beta = \frac{1}{V}\left(\frac{\partial V}{\partial P}\right)_T = \frac{k_2}{P}$$

or $\left(\frac{\partial V}{\partial P}\right)_T = \frac{-k_2 V}{P}$ (3)

Substituting equations (2) and (3) into equation (1) yields

$$dV = \frac{k_1 C_P T^{\,(C_P/C_V)^{-1}}\, V}{C_V}\, dT - \frac{k_2 V}{P}\, dP$$

or $$\frac{dV}{V} = k_1\frac{C_P}{C_V}T^{(C_P/C_V)-1}\ dT - k_2\frac{dP}{P}$$

Inteograting equation (4) gives

$$1n\ V = k_1\frac{C_P}{C_V}\cdot\frac{C_V}{C_P}\ T^{\ C_P/C_V} - k_2\ 1n\ P + \text{constant}$$

$$1n\ V = k_1 T^{\ C_P/C_V} - k_2\ 1n\ P + \text{constant}$$

$$V = e^{K_1 T^c P/C_v} - k_2\ P + \text{constant}$$

or $$P^{K_2}\ V = k\exp\left(k_1\ T^{\ (C_P/C_V)}\right)$$

Hence the desired equation of state is

$$P^{k_2}\ V = k\exp\left(k_1 T^{\ C_P/C_v}\right)$$

Solution 45. Expanding the given equation of state, P(V – *nb*) = nRT, gives PV – Pnb = nRt and rearranging to solve for V gives

$$V = \frac{nRT}{P} + nb \quad (1)$$

where V = volume, n = number of moles, P = pressure, R = gas constant and T = temperature. Differentiating equation (1) with respect to T and keeping P constant yields

$$\left(\frac{\partial V}{\partial T}\right)_P = \frac{nR}{P} \quad (2)$$

But $P = \dfrac{nRT}{v-nb}$. Therefore, $$\left(\frac{\partial V}{\partial T}\right)_P = \frac{nR}{\left[\dfrac{nRT}{V-nb}\right]} = \frac{nR}{nRT}\times V - nb$$

$$= \frac{V-nb}{T} \quad (2a)$$

Note that the subscript P signifies that the pressure is being held constant while equation (1) is being differentiated. Now, differentiating with respect to P and holding T constant gives

$$\left(\frac{\partial V}{\partial P}\right)_T = -\frac{nRT}{P_2} = -\left(\frac{nRT}{P}\right)\frac{1}{P} \qquad (3)$$

Observe that equation (1) can be rearranged to give V–nb = $\frac{nRT}{P}$

Therefore,

$$\left(\frac{\partial V}{\partial P}\right) = -\left(\frac{nRT}{P}\right)\frac{1}{P} = -(V - nb)\frac{1}{P} \qquad (3a)$$

Volume as a function of pressure and temperature can be represented as

$$V = f(P, T) \qquad (4)$$

If the volume is a property of the system for which P(V – *nb*) = *n*RT, then it has an exact differential

$$dV = \left(\frac{\partial V}{\partial P}\right)_T dP + \left(\frac{\partial V}{\partial T}\right)_P dT \qquad (5)$$

using equations (2a) and (3a), equation (5) becomes

$$dV = -\left(\frac{V-nb}{P}\right)dP + \left(\frac{V-nb}{T}\right)dT$$

$$= (V-nb)\left[-\frac{dP}{P} + \frac{dT}{T}\right]$$

Solution 46. Rearranging the equation P $\overline{V}$ = RT + APT – BP gives

$$P\overline{V} - APT + BP = RT$$

or

$$P(\overline{V} - AT + B) = RT.$$

Solving for P yields

$$P = \frac{RT}{\overline{V} - AT + B} \qquad (1)$$

$$= \frac{RT}{\overline{V}(1 - (AT/\overline{V}) + (B/\overline{V})}$$

$$P = \frac{RT}{\bar{V}(1 + (B - AT)/\bar{V}}$$

$$P\bar{V} = \frac{RT}{[1 + (B - AT)/\bar{V}]} \qquad (2)$$

Consider the term $|(B-AT)/\bar{V}|$ to be very much less than 1, then equation (2) can be rewritten as,

$$P\bar{V} = RT\,[1-(B-AT)/\bar{V}\,] \qquad (3)$$

The Boyle temperature of a gas according to one definition is the temperature, T_B, at which

$$\lim_{\bar{V} - \infty}\left[\frac{\partial (P\bar{V})}{\partial (1/\bar{V})}\right]_T = 0$$

at which point P = 0 and T = T_B. Thus, from equation (3),

$$B - AT_B = 0$$

or

$T_B = B/A$

(*b*) To be able to ascertain whether a gas has a critical point, the first and second derivative of the equation of state for such a gas with respect to V at constant T must vanish.

Therefore from (1)

$$\left(\frac{\partial P}{\partial V}\right)_T = \frac{-RT}{(\bar{V} - AT + B)^2} \qquad (4)$$

$$\left(\frac{\partial P}{\partial \bar{V}^2}\right)_T = \frac{2RT}{(\bar{V} - AT + B)^3} \qquad (5)$$

At the critical point $(\partial P/\partial \bar{V})_T = 0$ and $(\partial P/\partial \bar{V}^2)_T = 0$. The right hand sides of (4) and (5) can only be 0 if T = 0 or $\bar{V}$ – AT + B is inifinte. Therefore this gas does not have a true critical point.

Solution 47. It is not easy to rearrange the van der Waals equation into the virial form, which would allow T_B to be evaluated by inspection. Therefore, the van der Waals equation to be used is

$$P = \frac{nRT}{V-nb} - \frac{n^2a}{V^2} \qquad (1)$$

where P = pressure, V = volume, n = number of moles, T = temperature and R = gas constant. Multiplying both sides of equation (1) by V gives

$$PV = nRT\left(\frac{V}{V-nb}\right) - \frac{n^2a}{V} \qquad (2)$$

Differentiating equation (2) partially with respect to P, keeping T constant, gives

$$\left(\frac{\partial(PV)}{\partial P}\right)_T = \left[\frac{nRT}{V-nb} - \frac{nRTV}{(V-nb)^2} + \frac{n^2a}{V^2}\right]\left(\frac{\partial V}{\partial P}\right)_T \qquad (3)$$

At P = 0, $\left[\frac{\partial (PV)}{\partial P}\right]_T = 0$. Therefore,

$$\left[\frac{nRT}{V-nb} - \frac{nRTV}{(V-nb)^2} + \frac{n^2a}{V^2}\right]\left(\frac{\partial V}{\partial P}\right)_T = \qquad (4)$$

Under all condition $(\partial V/\partial P)_T < 0$; therefore, the expression in sqaure brackets in equation (4) must equal to zero. That is

$$\frac{nRT}{V-nb} - \frac{nRTV}{(V-nb)^2} + \frac{n^2a}{V^2} = 0 \qquad (5)$$

Multiplying equation (5) through by V^2 $(V-nb)^2$ gives

$$nRTV^2(V-nb) - nRTV^3 + n^2a(V-nb)^2 = 0$$

or

$$nRTV^2(V-nb) - nRTV^3 = -n^2a(V-nb)^2$$

Therefore,

$$RT = \frac{a}{b}\left[\frac{V-nb}{V}\right]^2 = \frac{a}{b}\left[\frac{V}{V} - \frac{nb}{V}\right]^2$$

$$= \frac{a}{b}\left[1 - \frac{nb}{V}\right]^2 \qquad (6)$$

At the Boyle point T = T_B, P = 0 and V = ∞, so RT_B = a/b or

$$T_B = \frac{a}{Rb}.$$

Another method of doing this is to set *n*/V = c. So, the van der Waals equation of state becomes

$$P = \frac{cRT}{1-bc} - ac^2 \qquad (7)$$

From equation (7),

$$PV = n\left[\frac{RT}{(1-bc)^2} - ac\right] \qquad (8)$$

Using the chain rule to differentiate equation (8) partially with respect to P gives

$$\left(\frac{\partial\,(PV)}{\partial P}\right)_T = \left(\frac{\partial (PV)}{\partial c}\right)_T - \left(\frac{\partial c}{\partial P}\right)_T = n\left[\frac{bRT}{(1-bc)^2} - a\right]\left(\frac{\partial c}{\partial P}\right)_T \qquad (9)$$

Observe that $\left(\frac{\partial c}{\partial P}\right) = \frac{1}{(\partial P/\partial c)_T} = \left(\frac{\partial P}{\partial c}\right)_T^{-1}$ From equation (7),

$$\left(\frac{\partial P}{\partial c}\right)_T = \left[\frac{bcRT}{(1-bc)^2} + \frac{RT}{1-bc} - 2ac\right]^{-1} \qquad (10)$$

Observe that equation (10) reduces to

$$\text{RT as P} \rightarrow 0 \text{ and c} \rightarrow 0 \text{ or } \left(\frac{\partial c}{\partial P}\right)_T = 1/\text{RT}.$$

Since 1/RT ≠ 0 and $n \neq 0$, the term in the square brackets in equation (9) must vanish at $c = 0$ and T T_B. That is,

$$bRT_B - a = 0 \qquad (11)$$

Solving for TB changes equation (11) to

$$T_B = \frac{a}{bR}$$

Solution 48. Looking at $\frac{P_cV_c}{RT_c} = \frac{(a/27b^2)(3b)}{(R)(8a/27Rb)}$ it is seen that T_c is proportional at a/b. If this ratio is computed for the various gases, then gas A, $a/b = 6/0.025 = 240$; For gas B, $a/b = 6/0.15 = 40$; for gas C, $a/b = 20/0.10 = 200$ and for gas D, $a/b = 0.05/0.02 = 2.5$ These values show that gas A (with the highest *a*/b ratio) has the highest T_c.

The constant *b* gives *a* measure volume (that is, $Vc = 3b$). Therefore, gas B has the largest molecules. At STP, the gas that has the lowest critical temperature and pressure will be most ideal. Gas D has the lowest a/b ratio, hence lowest Tc. The same gas also gives the lowest a/b^2 ratio, hence the lowest critical pressure P_c.

Solution 49. The van der Waals equation is

$$\left(P + \frac{n^2a}{V^2}\right)(V - nb) = nRT$$

where P = pressure

T = temperature

V = volume

n = number of moles.

a and *b* are van der Waals constants.

(*a*) It is known T_c, the critical temperature is proportional to the ratio a/b. Computing the a/b rations for the three gases yields :

for gas A $\frac{4.0}{0.027} = 148.15$

for gas B $\frac{12.0}{0.030} = 400$

for gas C $\frac{6.0}{0.032} = 187.5$

Since the ratio for gas B is the largest, gas B has the largest T_c. This question could have been answered by mere inspection. It can be seen that the constant *b* has almost the same value for the three gases. So, the larger a value means greater intermolecular forces. A larger T_c would be needed to overcome them. Since gas B has the largest a value, its T_c will be the largest.

(*b*) The constant *b* is proportional to V_c. Therefore gas C has the largest V_c.

(*c*) The *b* values are almost the same. Therefore ideality will be determined by the a values. Looking at the original equation, it can be seen that it is near ideality for values of the constants close to zero. Since gas A has the smallest a value, it should be the most ideal in behavior. Remember that for ideality the expression is PV = *n*RT.

Solution 50. Density $= \frac{\text{weight}}{\text{volume}} = \frac{W}{V}$

$$\frac{\text{density}}{\text{pressure}} = \frac{W/V}{P} = \frac{\rho}{P} \quad (1)$$

Where ρ = W/V.

$$\text{Compressibility factor} = \frac{PV}{RT} \qquad (2)$$

Therefore, comparing (1) and (2), it is observed that ρ/P is inversely proportional to the compressibility factor at a given temperature, so the maximum in ρ/P means a minimum in PV/RT. In plots of PV/RT vs. P/P_c, such a minimum occurs only for $T < T_c$ and $P > P_c$ where T_c and P_c and critical temperature and pressure, respectively. The data tells us that this minimum is between 1 atm and 20 atm. Therefore P_c could be greater or less than 1 atm and greater or less that 10 atm, but it must be less than 20 atm.

Solution 51.

$$P = \frac{RT}{(\overline{V}-b)} \cdot e^{-A/RT^{3/2}\overline{V}} \qquad (1)$$

Differentiate equation (1) twice with respect to $\overline{V}$ at constant temperature, T to generate two new equations. That is,

$$\left(\frac{\partial P}{\partial \overline{V}}\right)_T = \frac{RT}{(\overline{V}-b)}\, e^{-A/RT^{3/2}}\overline{V} \left(\frac{A}{RT^{3/2}\overline{V}^2} - \frac{1}{(\overline{V}-b)}\right) \tag{2}$$

and

$$\left(\frac{\partial^2 P}{\partial \overline{V}^2}\right)_T = \left\{\left[\frac{A}{RT^{3/2}\overline{V}^2} - \frac{1}{(\overline{V}-b)}\right]\left[\frac{-1}{(\overline{V}-b)^2} + \frac{A}{RT^{3/2}\overline{V}^2}\right] + \right.$$

$$\left. + \left(\frac{1}{(\overline{V}-b)}\right)\left[\frac{-2A}{RT^{3/2}V^{-3}} + \frac{1}{(\overline{V}-b)^2}\right] RTe^{-A/RT^{3/2}}\overline{V} \right. \tag{3}$$

The first and second derivatives disappear at the critical point, therefore,

$$\left(\frac{\partial P}{\partial \overline{V}}\right)_T = 0 \text{ and } \left(\frac{\partial^2 P}{\partial \overline{V}^2}\right)_T = 0$$

From equation (2), since $\frac{RT}{(\overline{V}-b)} e^{-A/RT^{3/2}}\overline{V}$ can not be zero, then

$$\frac{A}{RT_c^{3/2}\ \overline{V}_c^2} - \frac{1}{\overline{V}_c - b} = 0 \tag{4}$$

In equation (3), the first term is also zero; consequently, the second term will disappear only when

$$\frac{-2A}{RT_c^{3/2}\ V_b^3} + \frac{1}{\overline{V}_c - b^2} = 0 \tag{5}$$

Divinding equation (4) by equation (5) yields

$$\frac{2}{V_c} = \frac{1}{(\overline{V}_c - b)}$$

$$2(\overline{V}_c - b) = V_c$$

$$2\overline{V}_c - 2b = V_c$$

and $$\overline{V}_c = 2b \tag{6}$$

Substituting equation (6) into equation (4) gives

$$\frac{A}{RT_c^{3/2}\,(2b)^2} - \frac{1}{2b-b} = 0$$

$$\frac{A}{RT_c^{3/2}\,4b^2} - \frac{1}{b} = 0$$

$$RT_c^{3/2}\,4b^2 - Ab = 0$$

$$RT_c^{3/2}\,4b - A = 0$$

$$T_c^{3/2} = \frac{A}{4bR}$$

or $$T_c = \left(\frac{A}{4bR}\right)^{2/3} \tag{7}$$

Combining equations (6), (7), and (1) gives

$$P_c = \frac{R\left(\frac{A}{4bR}\right)^{2/3}}{(2b-b)}\; e^{-A/R(A/4bR)\,2b}$$

or $$P_c = \frac{R}{b}\left(\frac{A}{4bR}\right)^{2/3} e^{-2}$$

Solution 52. For a compressibility factor greater than unity, P = 500 atm and T = 500°C (that is the highest P and T) are chosen. The constant b = $V_c/3$. It is known that

$$\frac{P_cV_c}{RT_c} = \frac{\left(\frac{a}{27b^2}\right)(3b)}{R\left(\frac{8a}{27Rb}\right)} = \frac{3}{8}$$

for a van der Waals gas. Therefore, $V_c = \frac{3RT_c}{8P_c} = \frac{(3)\,(.082)\,(473^\circ K)}{(8)\,(30\text{ atm})}$ $T_c = 0.4848$ and $b = \frac{0.4848}{3} = 0.162$ liter/mole

Solution 53. $P = \frac{RT}{(\bar{V}-b)} - \frac{a}{T\bar{V}^2}$ (1)

Taking the first derivative with respect to v at constant temperature gives

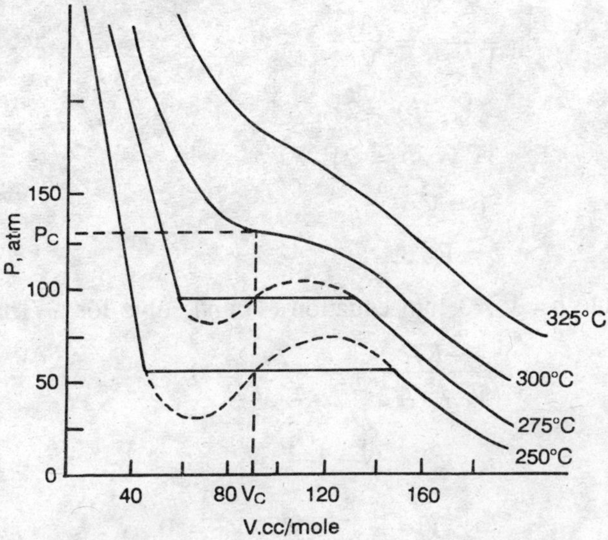

$$\left(\frac{\partial P}{\partial \overline{V}}\right)_T = \frac{-Rt}{(\overline{V}-b)^2} + \frac{2a}{T\overline{V}^3} \qquad (2)$$

The second derivative is

$$\left(\frac{\partial^2 P}{\partial \overline{V}}\right)_T = \frac{RT}{(\overline{V}-b)^3} - \frac{6a}{T\overline{V}^4} \qquad (3)$$

Applying the critical isotherm conditions, that the partial derivatives with respect to volume at constant temperature is zero, equations (2) and (3) become

$$\left(\frac{\partial P}{\partial \overline{V}}\right)_T = \frac{-RT}{(\overline{V}-b)^2} + \frac{2a}{T\overline{V}^3} = 0 \qquad (4)$$

and

$$\left(\frac{\partial^2 P}{\partial \overline{V}^2}\right)_T = \frac{2RT}{(\overline{V}-b)^3} - \frac{6a}{T\overline{V}^4} = 0 \qquad (5)$$

Solving equations (4) and (5) simulataneously yields

$$\frac{2}{\overline{V}_c - b} = \frac{3}{\overline{V}_c}$$

$$3(\overline{V}_c - b) = 2\overline{V}_c$$

$$3\overline{V}_c - 3b = 2\overline{V}_c$$

$$3b = \overline{V}_c$$

$$b = \overline{V}_c/3.$$

Substitute $b = \overline{V}_c/3$ into equation (4) and solve for a. That is,

$$\frac{-RTc}{\left(\overline{V}_c - \overline{V}_c/3\right)^2} + \frac{2a}{T_c\,\overline{V}_c^3} = 0$$

or

$$2a\overline{V}_c^2\left(1 - \tfrac{1}{3}\right)^2 - RT_c^2\,\overline{V}_c^3 = 0$$

$$2a = \frac{RT_c^2\ \overline{V}_c^3}{\frac{4}{9}\overline{V}^2}$$

$$2a = \frac{9RT_c^2\ \overline{V}_c}{4}$$

$$a = \frac{9RT_c^2\,\overline{V}_c}{8} \qquad (5)$$

Substituting $a = 9RT_c^2\ \overline{V}_c/8$ and $b = \overline{V}_c/3$ into equation (1) gives

$$P_c = \frac{RT_c}{\left(\overline{V}_c - \overline{V}_c/3\right)} - \frac{\frac{9RT_c^2\,\overline{V}_c}{8}}{(T_c)\,(\overline{V}_c^2)}$$

$$= \frac{RT_c}{\overline{V}_c(1 - \frac{1}{3})} - \frac{9RT_c^2\,\overline{V}_c}{8T_c\,\overline{V}_c^2}$$

$$= \frac{RT_c}{\frac{2}{3}} \overline{V}_c - \frac{9RT_c}{8\overline{V}_c}$$

$$P_c = \frac{3RT_c}{2\overline{V}_c} - \frac{9RT_c}{8\overline{V}_c}$$

$$= \frac{12RT_c - 9RT_c}{8\overline{V}_c}$$

$$P_c = \frac{3Rt_c}{8\overline{V}_c}$$

and $$R = \frac{8P_c \overline{V}_c}{3T_c}$$

Substituting R = $8P_c \overline{V}_c/3T_c$ into (5) gives

$$a = \frac{9(8P_c V_c)\, T_c^2\, \overline{V}_c}{24T_c}$$

or $$a = 3P_c\overline{V}_c^2 T_c$$

Now, define the reduced variables as π, φ and θ, corresponding to reduced pressure, volume and temperature respectively.

$$\pi = \frac{P}{P_c} \tag{6}$$

$$\varphi = \frac{\overline{V}}{\overline{V}_c} \tag{7}$$

and $$\theta = \frac{T}{T_c} \tag{8}$$

From equation (6) P = πP_c and from equations (7) and (8)

$\overline{V} = \varphi\overline{V}_c$ and $T = \theta T_c$ respectively.

Substituting $P = \pi P_c$, $\overline{V} = \varphi\overline{V}_c$, $T = \theta T_c$, R, b and a into equation (1) gives

$$\pi P_c = \frac{\frac{8P_c V_c}{3T_c}(\theta T_c)}{\varphi V_c - V_c/3} - \frac{3P_c\overline{V}_c^2 T_c}{\theta T_c \varphi \overline{V}^2}$$

$$= \frac{8P_c V_c \theta T_c}{3T_c V_c(\varphi - \frac{1}{3})} - \frac{3P_c}{\theta\varphi^2}$$

$$\pi P_c = \frac{8P_c\theta}{3(\varphi - \frac{1}{3})} - \frac{3P_c}{\pi\,\varphi^2}$$

or

$$\pi = \frac{8\theta}{3\varphi - 1}\ \frac{3}{\theta\,\varphi^2} \tag{9}$$

Equation (9) is the equation of state in terms of reduced variables.

Solution 54. Since this problem requires approximate values, it is not necessary to interpolate or extrapolate. For the purposes of the problem, the 300°C isotherm (which is the closest to the critical point) is good enough.

Isotherm, means, for any pressure and volume on that line, the temperature remains the same (300°C in this case). The terms P_c and V_c can be estimated from the point of inflection. Therefore $P_c \approx 130$ atm, $V_c \approx 90$ cm^3/mole and $T_c = 300$°C $= 573$°K.

For a van der Waals gas, $V_c = 3b$ and $P_c = \dfrac{a}{27b^2}$.

A system of two equations and two unknows can now be presented.

Thus,

$$90 = 3b \tag{1}$$

and

$$130 = a = \frac{a}{27b^2} \tag{2}$$

From (1) b = 90/3 = 30 cm^3/mole. Substituting this value into (2),

$$130 = \frac{a}{27 \times 30^2}$$

$$a = (130 \times 27 \times 30^2) \;\; cc^2 \text{ atm/mole}^2$$
$$= 3159000 \; cc^2 \text{ atm/moles}^2$$
$$= 3.16 \times 10^6 \; cc^2 \text{ atm/moles}^2 .$$

Solution 55. Rearranging equation (1) to solve for P gives

$$P = \frac{nRT}{V - nb'} e^{na'/RTV}$$

For 1 mole,

$$P = \frac{RT}{V - b'} e^{a'/RTV} \qquad (2)$$

where P = pressure, R = gas constant, V = volume and T = temperature. Differentiating equation (2) with respect with V gives

$$\frac{dP}{dV} = -\frac{P}{V - b'} + \frac{a'P}{RTV^2} \quad (3)$$

and

$$\frac{d^2P}{dV^2} = \frac{a'^2P}{(RTV^2)^2} - \frac{2a'P}{RTV^3} \qquad (4)$$

At some point on a P–V diagram, the curve is horizontal and this part can be mathematically expressed as dP/dV = 0. Also, the highest temperature (maximum) at which such a horizontal line can still exist is called the critical temperature and at this point.

$$\frac{d^2P}{dV^2} = 0$$

So, using these two conditions for equations (3) and (4), they can be rewritten as

$$0 = -\frac{P}{V-b'} + \frac{a'P}{RT^2} \qquad (3a)$$

$$0 = -\frac{a'^2P}{RTV} - \frac{2a'P}{RTV^3} \qquad (4a)$$

Since there are two equations and two unknows, the equations can be solved simultaneously. Simplfying equation (4a), it becomes

$$a'\left(\frac{a'P}{R^2T^2V^4} - \frac{2P}{RTV^3}\right) = 0$$

From this, observe that there are two possible values for a′. That is, a′ = 0 and a′ = $2RT_cV_c$. The trivial solution is) discarded, so that $a' = 2RT_cV_c$. Putting this into equation (3a) gives

$$0 = -\frac{P}{V_c - b'} + \frac{2RT_cV_cP}{RT_cV_c^2}$$

or

$$0 = -\frac{P}{V_c - b'} + \frac{2P}{V_c} = P\left(\frac{-1}{V_c - b'} + \frac{2}{V_c}\right)$$

From this, $-\dfrac{1}{V_c - b'} + \dfrac{2}{V_c} = 0$ or

$$\frac{1}{V_c - b'} = \frac{2}{V_c} \qquad (5)$$

The reciprocal of equation (5) is

$$V_c - b' = \frac{V_c}{2},$$

or

$$-b' = \frac{V_c}{2} - V_c = -\frac{V_c}{2}.$$

$$b' = \frac{V_c}{2}.$$

Solution 56. For any gas that obeys van der Waals equation, $V_c = 3b$ $P_c = a/27b^2$ and $T_c = 8a/27bR$.

(*a*) If we take the ratio T_c/P_c, we have

$$\left[\frac{8a}{27bR}\right] \times \left[\frac{27b^2}{a}\right] = \frac{8b}{R}$$

Therefore b is directly proportional to T_c/P_c. Also it is observed that $V_c = 3b$ is proportional to T_c/P_c .

T_c/P_c for NO gas = 177/64 = 2.766.

T_c/P_c for CCl_4 = 550/45 = 12.22.

Therefore NO should have the smaller b value.

(*b*) The constant a is proportional to $P_cV_c^2$ or to T_c^2/P_c. Therefore it will be smaller for NO.

(*c*) NO should be more ideal since 300°K is above its critical temperature, but below that of CCl_4.

Solution 57. The van der Waals equation is

$$\left(P + \frac{n^2a}{V^2}\right)(V - nb) = nRT$$

(*a*) For the compressibility factor to be greater than unity, what is needed is a high pressure and temperature. By this it is meant that the required temperature and pressure must both be greater than the values at the critical point. Hence 120°C and 500 atm is the correct answer.

(b) Compressibility factor at the critical point = $\dfrac{P_cV_c}{RT_c}$,

where the subscript *c* means the properties at the critical point. For a gas that obeys the van der Waals equation,

$$T_c = \frac{8a}{27Rb},$$

$V_c = 3b$ and $P_c = \dfrac{a}{27b^2}$ where *a* and *b* are constants (as in equations (1).

Now after substituting these values into $\dfrac{P_cV_c}{RT_c}$, it yields

$$\frac{\left(\frac{a}{27b^2}\right)(3b)}{(R)\left(8\frac{a}{27Rb}\right)}$$

$$\text{Compressibility factor} = \frac{3ab}{27b^2} \times \frac{27Rb}{8aR} = \frac{3}{8}.$$

Solution 58. $\left(P + \frac{n}{V^2 T^{1/2}}\right)(V - m) = RT$

solving for P gives

$$P = \frac{RT}{V - m} - \frac{n}{V^2 T^{1/2}} \quad (1)$$

Differentiate equation (2) twice with respect to V at constant T and set them equal to zero because at the critical point

$$\left(\frac{\partial P}{\partial V}\right)_T = \left(\frac{\partial^2 p}{\partial V^2}\right)_T = 0$$

Therefore at the critical point

$$\left(\frac{\partial P}{\partial V}\right)_T = \frac{RT_c}{(V_c - m)^2} + 2\frac{n}{V_c^3 T_c^{1/2}} = 0 \qquad (2)$$

$$\left(\frac{\partial^2 P}{\partial V^2}\right)_T = \frac{2RT_c}{(V_c - m)^3} - 6\frac{n}{V_c^4 T_c^{1/2}} = 0 \qquad (3)$$

Equation (2) can also be written as

$$\frac{RT_c}{V_c - m)^2} = \frac{n}{T_c^{1/2}} \frac{2}{V_c^3} \qquad (4)$$

and equation (3) as

$$\frac{RT_c}{(V_c - m)^3} = \frac{n}{T_c^{1/2}} \frac{3}{V_c^4} \qquad (5)$$

Divide equation (4) by equation (5) to get

$$V_c - m = \frac{2}{3} V_c$$

from which $m = \dfrac{1}{3} V_c$

where V_c is the critical volume. Substitue the value of m into equation (4) and solve for *n*.

$$\frac{RT_c}{(V_c^2 - 2mV_c + m^2} = \frac{n}{T_c^{1/2}} \frac{2}{V_c^3}$$

or $$n = \frac{RT_c^{3/2} V_c^3}{2V_c^2 - \frac{4}{3}V_c^2 + \frac{2}{9}V_c^2}$$

$$P_c = \frac{RT_c}{V_c - m} - \frac{9/8\, RT^{3/2} V_c}{V_c^2 T_c^{1/2}}$$

$$= \frac{3}{2} \frac{RT_c}{V_c} - \frac{9}{8} \frac{RT_c}{V_c}$$

$$= \frac{3}{8} \frac{RT_c}{V_c}$$

or

$$\frac{P_c V_c}{RT_c} = \frac{3}{8} \qquad (7)$$

CO_2 has a molecular weight of 44g mol^{-1}

$$V = \left(6.948 \frac{(dm)^3}{kg}\right)\left(\frac{10^3 cm^3}{(dm)^3}\right)\left(\frac{0.044kg}{mol}\right)$$

$$= 3.057 \times 10^2\ cm^3/\ mol.$$

From equation (7) $V_c = \dfrac{3}{8} \dfrac{RT_c}{P_c}$

$$= \frac{3}{8} \times \frac{(82.05cm^3\ atm\ /K^{-1}\ mol^{-1})\ (304.2°K)}{72.9atm}$$

$$= 128.3\ cm^3\ mol^{-1}$$

Remember that m = $V_c/3$

$$= \frac{128.3 \text{cm}^3 \text{ mol}^{-1}}{3}$$

$$= 42.79 \text{ cm}^3 \text{ mol}^{-1}$$

From equation (6

$$n = \frac{(9)\,(82.05\text{cm}^3 \text{ atm}^\circ\text{K}^{-1} \text{ mol}^{-1})\,(304.2^\circ\text{K})^{3/2}\,(128.3\text{cm}^3 \text{ mol}^{-1}}{8}$$

$$= 6.28 \times 10^7 \frac{\text{cm}^6 \text{ atm K}^{1/2}}{\text{mol}^2}$$

Solving for P using equation (1) gives

$$P = \left[\frac{(82.05 \text{ cm}^3 \text{ atm K}^{-1} \text{ mol}^{-1})\,(373^\circ\text{K})}{(3.057\times 10^2 \text{ cm}^3/\text{mol}) - (42.79 \text{ cm}^2/\text{mol})}\right]$$

$$- \left[\frac{6.28\times 10^7 \, cm^6 \, atmK^{1/2}/mol^2}{(3.057\times 10^2 \, cm^3/mol)^2\,(373^\circ K)^{1/2}}\right]$$

$$= 81.57 \text{ atm.}$$

By defintion the compressibility factor z is given as

$$Z = \frac{PV}{RT}$$

$$= \frac{(81.57 \text{ atm})\,(3.057\times 10^2\text{cm}^3/\text{mol})}{(82.05\text{cm}^3 \text{ atm}^\circ\text{K}^{-1} \text{ mot}^{-1})(373^\circ\text{K})}$$

$$= 0.815.$$

Solution 59. $\frac{P_1V_1}{T_1} = \frac{P_2V_2}{T_2}$; $V = \frac{4}{3}\pi\left(\frac{d}{2}\right)^3$;

$$\frac{3\times(4\pi/3)\,(1/2)^3}{278} = \frac{1\times(4\pi/3)\,(d/2)^3}{298};$$

$$\frac{3}{278} = \frac{d^3}{298};\ d^3 = 3.22;\ d = 1.48 \text{ cm}$$

Solution 60. Weight of ozone = $W = 6.7624 - 6.5998 = 0.1626$ g

$$M = \frac{WRT}{PV} = \frac{0.1626 \times 82.06 \times 301.4}{(274.4/760.0) \times 235.67} = 47.3 \text{ g mole}^{-1}$$

Solution 61. $M = \frac{WRT}{PV} = \frac{0.1023 \times 82.06\ 305.7}{(743.95/760) \times 35.33} = 74.19$ g mole^{-1}

Making a correction for relative humility, Mac Innes and Kreiling found 75.72. The value related from atomic weights is 74.12.

Solution 62.

P	0.400	0.8000	1.0000
d/P	3.780	3.860	3.900 g liter^{-1} atm^{-1}

For an ideal gas, PM = dRT, or M = dRT/P. For a real gas, M = RT $\lim_{P \to 0}$ (d/P). We plot d/P as a function P and extrapolate to P = 0. From the graph we see that $\lim_{p \to 0}$ (d/P) = 3.700 g liter^{-1} atm^{-1}

M = 0.08206 liter atm °K^{-1} mole^{-1} × 300.0°K × 3.700 g liter^{-1} atm^{-1} = 91.09g mole^{-1}

Solution 63. Plot (d/P) vs P and extrpolate $P = 0$.

P, atm	0.2	0.5	0.8
d/P, g liter^{-1} atm^{-1}	1.398	1.416	1.4344

$\lim_{P \to 0}$ (d/P) = 1.3857 (from graph)

$$M = \left(\frac{d}{P}\right)_0 RT = 1.3857 \times 0.08206 \times 273.15 = 31.07$$

(Theoretical for $CH_3\,NH_2$ = 31.06)

Solution 64.

$$\text{Fraction} = \frac{\int_{\varepsilon 0}^{\infty} e^{-\varepsilon/kT}\sqrt{\varepsilon}\, d\varepsilon}{\int_{0}^{\infty} e^{-\varepsilon/kT}\sqrt{\varepsilon}\, d\varepsilon} = \frac{2}{\pi^{1/2}(kT)^{3/2}} \int_{\varepsilon 0}^{\infty} e^{-\varepsilon/kT}\sqrt{\varepsilon}\, d\varepsilon$$

Solution 65.

$$\frac{RT}{N} 1\text{n} \frac{n_1}{n_2} = \frac{4}{3}\pi r^3 g\,(h_2 - h_1)\,(d - d')$$

$$\frac{8.31 \times 10^7 \times 288.2 \times 2.303}{N} \log \frac{100}{47}$$

$$\times \frac{4}{3}\pi\,(2.12\ 10^{-5})^3 \times 981 \times (35 - 5) \times 10^{-4} \times (1.206 - 0.999)$$

N = 7.44×10^{23}

(Note : Use R in ergs $^{\circ}K^{-1}$ mole^{-1})

Solution 66. The maximum will occur at the most probable velocity,

$$\upsilon_p = \left(\frac{2RT}{M}\right)^{1/2} = \left(\frac{2 \times 8.314 \times 10^7 \times 273.2}{2.016}\right)^{1/2}$$

$$= 1.50 \text{ ö } 10^5 \text{cm sec}^{-1}$$

At this velocity

$$f(\upsilon) = 4\pi \upsilon^2 \left(\frac{M}{2\pi RT}\right)^{3/2} e^{-Mv^2/2RT}$$

$$= 4\pi\,(1.50 \times 10^5)^2 \left(\frac{2.016}{2\,\pi \times 8.314 \times 10^7 \times 273.2}\right)^{3/2}$$

$$\times \exp\left(\frac{-2.016\,(1.51 \times 10^5)^2}{2 \times 8.314 \times 10^2 \times 273.2}\right)$$

$$f(\upsilon) = 5.54 \times 10^{-6}$$

The values of $f(\upsilon)$ at various υ's, calculated in a similar manner, are given below :

$\upsilon \times 10^{-5}$	0.50	1.00	1.50	2.00	3.00	4.00 cm sec–1
$f(\upsilon) \times 10^{6}$	1.49	4.28	5.54	4.53	1.11	0.89

The graph is shown

Solution 67. $C = \dfrac{PN}{RT} = \dfrac{(1/760) \times 1.013 \times 10^6 \times 6.023 \times 10^{23}}{8.314 \times 10^7 \times 298.2}$

$= 3.24 \times 10\ {}^{1}6$ molecules cm^{-2}

$$l = \frac{1}{\sqrt{2}\pi\sigma^2 C} = \frac{1}{\sqrt{2}\pi\,(2.86 \times 10^{-8})^2 \times 3.24 \times 10^{16}}$$

$= 8.49 \times 10^{-3}$ cm

Solution 68.

$$C = \frac{PN \times 10^{-3}}{RT} = \frac{(1/760) \times 1.013 \times 10^6 \times 6.023 \times 10^{23} \times 10^{-3}}{8.314 \times 10^{-7} \times 298.2}$$

3.24×10^{13} molecules cm^{-3}

$$\bar{\upsilon} = \left(\frac{8RT}{\pi M}\right)^{1/2} = \left(\frac{8 \times 8.314 \times 10^7 \times 298.2}{\pi \times 28.03}\right)^{1/2}$$

$= 4.75 \times 10^4$ cm sec^{-1}

$$Z = \frac{\sqrt{2}}{2}\pi\sigma^2\bar{\upsilon}C^2$$

$$= \frac{\sqrt{2}}{2}\pi\,(3.16 \times 10^{-8})^2 \times (4.75 \times 10^4) \times (3.24 \times 10^{13})^2$$

$= 1.11 \times 10^{17}$ collisions $cm^{-3}\ sec^{-1}$

Solution 69. The effusion rate of $H_2 = dn_{H_2}/dT = n_{H_2}\bar{\upsilon}_{H_2}A$ and the effusion rate of $D_2 = dn_{D_2}/dT = n_{D_2}\upsilon\bar{V}_{D_2}A$, or

$$\frac{dn_{H_2}/dt}{dn_{D_2}/dt} = \frac{n_{H_2}\bar{\upsilon}_{H_2}A}{n_{D_2}\bar{\upsilon}_{D_2}A} = \frac{n_{H_2}}{n_{D_2}} \cdot \frac{\bar{\upsilon}_{H_2}}{\bar{\upsilon}_{D_2}} = \frac{90}{10} \cdot \frac{\bar{\upsilon}_{H_2}}{\bar{\upsilon}_{D_2}} .$$

Graham's law contains the root-mean-square velocities but since these are proportional to the arithmetical mean velocities and the proportionality constant cancels out we may write

$$\frac{dn_{H_2}/dt}{dn_{D_2}/dt} = \frac{90}{10}\sqrt{\frac{M_{D_2}}{M_{H_2}}} = \frac{90}{10}\sqrt{\frac{4.028}{2.016}} = 9\sqrt{2.00} = 12.7.$$

Thus the relative effusion rate is 12.7 molecules H_2 per molecule D_2. Therefore the per cent composition of the initial gas passing through will be

$$\text{mole per cent } D_2 = \frac{1 \times 100}{12.7 + 1} = 7.30$$

$$\text{mole per cent } H_2 = 100 - 7.30 = 92.7$$

The temperature, pressure, and area data are extraneous.

Solution. 70.

$$(a)\ P = \frac{nRT}{V} = \frac{2 \times 0.08206 \times 298.2}{10.0} = 490 \text{ atm}$$

$$(b)\ P = \frac{nRT}{V - nb} - \frac{n^2 a}{V^2}$$

$$= \frac{2 \times 0.08206 \times 298.2}{10.0 - 2(0.1453)} - \frac{2^2 \times 25.43}{10.0^2}$$

$$= 4.02 \text{ atm}$$

Solution 71.

$$(a)\ V = \frac{nRT}{P} = \frac{0.1 \times 0.08206 \times 273.2}{100} = 22.4 \times 10^{-3} \text{ liter}$$

$$(b)\ V = \frac{ZnRT}{P} = \frac{0.2007 \times 0.1 \times 0.08206 \times 273.2}{100}$$

$$= 4.50 \times 10^{-3} \text{ liter}$$

Solution 72.

$$V_2 = \frac{P_1 V_1}{P_2} \cdot \frac{Z_2 T_2}{Z_1 T_1} = \frac{800 \times 1.00 \times 1.10 \times 373.2}{200 \times 1.95 \times 223.2} = 3.77 \text{ liters}$$

Solution 73.

$$Z_c = \frac{P_c V_c}{RT_c} = \frac{a}{27b^2} \cdot \frac{3b}{R} \cdot \frac{27bR}{8} = \frac{3}{8} = 0.375$$

Solution 74. It is approximately true for many substances that the normal boiling point in °K is two-thrids of the critical temperature in °K. Thus

$$\tfrac{2}{3}\, T_c = T_b = (68.9 + 273.2) = 342.1$$

$$T_c = 513°\text{K (accepted, 507.9°K) or } 240°\text{C}$$

Solution 75. Plot α vs P ; α at P = 0 ισ 366.9 × 10^{-6} (from graph). Gay Lussac's law; $V_t = V_0 (1 + \alpha\, t)$, where $\alpha = (1/V_0)\,(\partial V/\partial t)_P$. When P → 0, the behaviour of the gas approaches ideality and Gay-Lussac's law is obeyed. Thereforeuse ∝ at $P = 0$. When t = t_0, V_t should be zero, and under these conditions.

$$1 + \alpha\, t_0 = 0,\ t_0 = -\frac{1}{\alpha} = -\frac{1}{3660.9 \times 10^{-6}} = -273.16$$

Solution 76. $M = RT \lim\limits_{P \to 0} \left(\frac{\rho}{P}\right)$

$$\frac{\rho}{P} = (2.000 + 0.0200P)\ \text{g (liter at,)}^{-1}$$

$$\text{M} = \text{RT}(2.000)$$

$$= 0.082056 \text{ liter atm deg}^{-1} \text{ mole}^{-1} \times 300.00° \times 2.000\text{g (liter atm)}^{-1}$$

$$= 49.23 \text{ g mole}^{-1}$$

Solution 77. Applying the ideal gas law we find that at 0° C and 760 mm Hg (pressure not yet corrected) the weight of 1.000 liter of HC*l* is, for example,

$$\frac{1000}{465.856} \times \frac{760}{756.76} \times 0.76097 = 1.64053 \text{ g}$$

The average for all four sets of data is 1.64016 g liter^{-1} . Since the pressure was measured in London where the force of gravity is 1.000588 times that at 45° latitude, the true presures were $P_{observed} \times$ 1.000588 and the corrected average weight per liter is

$$\frac{1.64016}{1.000588} = 1.63919 \text{ g liter}^{-1}$$

The plotting PV vs P we find that at P = 0, the PV product is P_0V_0 = 55213; and at P = 760 torr, PV = P_1V_1 = 54803. Let y be the density of the HCl at 760 torr and 0°C. Then

$$\frac{P_0V_0}{P_1V_1} = \frac{55213}{54803} = \frac{1.63919}{y}$$

and

$$y = 1.62702 \text{ g liter}^{-1}$$

Now let x be the formula weight of HC*l*. Then

$$\frac{1.62702}{1.42762} = \frac{x}{31.9988} \text{ and } x = 36.467$$

Then the atomic weight of C1 is

$$36.467 - 1.008 = 35.459$$

This is based on the 1961 atomic weight scale. The accepted atomic weight of C1 on this scale is 35.453. Gray and Burt, using an older atomic weight scale, obtain a slightly different answer.

Solution 78. $V_{50} - V_0 = \int_0^{50} \propto V_0 \, dt$

$$= (1.817\times 10^{-4}t + 2.95\times^{-9}t^2 + 10^{-10}t^3]_0^{100} \ V_0$$

$$= (1.817\times 10^{-4}t + 2.95\times 10^{-9}t^2 + 1.15\times 10^{-10}t^3]_0^{100} \ V_0$$

$$= 0.009107 \ V_0$$

$$V_{100} - V_0 = \int_0^{100} \alpha \, V_0 \, dt$$

$$= (1.817\times 10^{-4}\,t + 2.95\times 10^{-9}\,t^2 + 1.15\times 10^{-10}\,t^5]_0^{100} \ V_0$$

$$= 0.01831\ V_0$$

$$t_{Hg} = \frac{0.009107V_0}{0.01831V_0} \times 100 = 49.8^\circ.$$

Solution 79. (*a*) This definition is based on the special properties of Hg; a Hg thermometer that is calibrated to agree with an ideal gas thermometer at two points will not agree at other points. The international scale, being based on the properties of any ideal gas (more precisely, any real gas in the limit ot zero pressure), is more satisfactory than a scale that singles out a preferred thermometric medium. Also this scale has as fixed points the freezing and boiling points of water (as the international scale had before 1954); temperatures near 0°K can be specified more reproducible if absolute zero is taken as one of the fixed points.

(*b*) This definition makes *T* proportional to the volume of O_2, a nonideal gas, at a fixed pressure. The international scale defines *T* by $T = 1/nR \lim_{i=0} (PV)$ for any gas, and thus avoids any dependence on the peculiarities of a specific gas. The ideal gas scale, thus defined, has the decisive advantage of being identical with the thermodynamic scale.

Solution 80. Let

W_{Ar} = the weight of argon in grams

W_{He} = the weight of helium in grams

Then

$$W_{Ar} + W_{He} = 5.00$$

Let

P = the total pressure = 1.00atm

P_{Ar} = the partial pressure of argon

P_{He} = the partial pressure of helium

Then

$$P = P_{Ar} + P_{He}$$

$$= n_{Ar}\frac{RT}{V} + n_{He}\frac{RT}{V} = \frac{W_{Ar}RT}{M_{Ar}V} + \frac{W_{He}RT}{M_{He}V}$$

$$= \left(\frac{W_{He}}{4.00} + \frac{5.00 - W_{He}}{39.9}\right) \times \frac{0.0821 \times 298}{10.0}$$

$$= 1.00$$

$W_{He} = 1.26g$

Weight per cent He = $100 \times \frac{1.26}{5.00} = 25.2$

Weight per cent Ar = 100 – 25.2 = 74.8

Solution 81. n_{H_2O} in RT liters saturated air = PV/RT = 0.0312 mole. RT = 24.47 liter atm per mole. The total number of moles of N_2 + O_2 +Ar remaining in 24.47 liters saturated air = 1– 0.0312 = 0.9688. Therefore the total weight of all gases plus water vapor in 24.47 liters saturated air = 2 × 14.0067 × 0.781 × 0.9688 + 2 × 15.9994 × 0.210 × 0.9688 + 39.948 × 0.009 × 0.9688 + 0.0312 × 18.02 = 21.18 + 6.51 + 0.35 + 0.56 = 28.60 g per 24.47 liters. The density of water-saturated air at 25°C and 1 atm total pressure is therefore

$$\frac{28.60}{24.47} = 1.169g\ liter^{-1}$$

Solution 82. (*a*)

$$\ln\frac{P_0}{P} = \frac{Mg(h - h_0)}{RT}$$

$$\log\frac{1}{P} = \frac{29 \times 980.6\,(1248 \times 12 \times 2.54 - 0)}{2.303 \times 8.314 \times 10^7 \times 298.2}$$

$$\log P = -0.0190 = 9.9810 - 10$$

$$P = 0.957 \text{ atm}$$

(*b*) Plot log *P* vs *h*. The plot must pass through the point log *P* = log 1 = 0 (if P is in atm), *h* = 0, with a slope of *Mg*/2.303RT.

Solution 83. $-dP = \rho g\, dz$ (where ρ = density)

$$= \frac{PMg}{RT}\, dz$$

$$\frac{dP}{P} = \frac{-Mg\, dz}{RT}$$

$$\ln\left(\frac{P}{P_0}\right) = \frac{-Mgz}{RT}$$

$$P = P_0\, e^{-Mgz/RT}$$

Solution 84.

$$\frac{dP}{P} = \frac{-Mg\, dz}{RT}\,;\; g = \frac{Gm_e}{r^2}$$

$$\frac{dP}{P} = \frac{-MgGm_E\, dz}{RTr^2}\,;\; z = r - r_e$$

where r_e is the radius of the earth.

$$\frac{dP}{P} = \frac{-MgGm_E\, dz}{RT\,(z + r_E)^2}$$

$$\ln\left(\frac{P}{P_0}\right) = \frac{MgGm_E}{RT}\left(\frac{1}{z + r_E} - \frac{1}{r_E}\right)$$

Solution 85. $\frac{dP}{P} = \frac{-Mg\, dz}{RT}\,;\; \frac{dP}{P} = \frac{-Mg\, dz}{R\,(T_0 - az)}$

Let $(T_0 - az) = x$. Then $dx = -a\, dz$, and $dz = -\, dx/a$.

$$\int_{x_0}^{x} -\frac{dx}{ax} = -\frac{1}{a}\ln\frac{x}{x_0} = +\frac{1}{a}\ln\left(\frac{T_0}{T_0 - az}\right)$$

$$\ln\left(\frac{P}{P_0}\right) = \frac{-Mg}{Ra}\ln\left(\frac{T_0 - az}{T_0}\right)$$

Solution 86. (*a*) The number of molecules having speeds between υ and $\upsilon + d\upsilon$ is $Ce^{-m\upsilon^2/2kT}\upsilon^2\, d\upsilon$, where C is some constant; this is proportional to the Boltzmann factor and to the volume $4\pi\upsilon^2\, d\upsilon$ of the region in velocity space between two concentric spheres of radi υ and $\upsilon + d\upsilon$. The number of molecules having speeds greater than υ_0 is $C\int_{r_0}^{\infty} e^{-m\upsilon^2/2kT}\upsilon^2\, d\upsilon$.

The total number of molecules is

$$C\int_0^{\infty} e^{-m\upsilon^2/2kT\upsilon^2} d\upsilon = C\frac{\sqrt{\pi}}{4}\left(\frac{2kT}{m}\right)^{3/2}$$

The fraction having $\upsilon > \upsilon_0$ is

$$\frac{\int_{\upsilon_0}^{\infty} e^{-m\upsilon^2/2kT}\upsilon^2\, d\upsilon}{\int_0^{\infty} e^{-m\upsilon^2/2kT}\upsilon^2\, d\upsilon} = \frac{4}{\sqrt{\pi}}\left(\frac{m}{2kT}\right)^{3/2}\int_{\upsilon_0}^{\infty} e^{-m\upsilon^2/2kT}\upsilon^2\, d\upsilon$$

This integral cannot be expressed as a finite combination of elementary functions.

(b) In two dimensions the number of molecules having speeds between υ and $\upsilon + d\upsilon$ is $Ce^{-m\upsilon^2/2kT}\upsilon\, d\upsilon$; we are now considering the region between circles of radi υ and $\upsilon + d\upsilon$ and the area of this region is $2\pi\upsilon\, d\upsilon$. The fraction having $\upsilon > \upsilon_0$ is

$$\frac{\int_{\upsilon_0}^{\infty} e^{-m\upsilon^2/kT}\upsilon\, d\upsilon}{\int_0^{\infty} e^{-m\upsilon^2/2kT}\upsilon\, d\upsilon} = e^{-m\upsilon_0^2/2kT}$$

Solution 87.

$$|\overline{\upsilon x}| = \frac{\int_0^{\infty} \upsilon_x e^{-m\upsilon_x^2/2kT}\, d\upsilon_x}{\int_0^{\infty} e^{-m\upsilon_x^2/2kT}\, d\upsilon_x} = \frac{\sqrt{2kT}}{\pi m} = \frac{\sqrt{2RT}}{\pi M}$$

Z coll $cm^{-2}\ sec^{-1} = \frac{1}{2}N$ molecules $cm^{-3} \times \sqrt{2RT/\pi M}$ cm sec–1, where $\frac{1}{2}$ is the fraction of moleculus moving toward the surface, and 2RT/M is the volume of the cylinder, 1 cm^2 in cross section, the molecules in which will strike the surface in 1 sec (if they are moving in the proper sense, that is, toward the surface),

$$Z = \frac{\frac{1}{2} \times 1 \text{ atm} \times 6.023 \times 10^{23} \text{ molecules mole}^{-1}}{82.06 \text{cm}^3 \text{atmdeg}^{-1} \text{mole}^{-1} \times 298 \text{deg}}$$

$$\times \frac{\sqrt{2} \times 8.314 \times 10^7 \times 298}{\pi} \left(\frac{0.2}{\sqrt{32}} + \frac{0.8}{\sqrt{28}} \right)$$

$$= 2.885 \times 10^{23} \text{ collisions cm}^{-2} \text{ sec}^{-1}$$

$$A = 2.0 \times 10^4 \text{cm}^2$$

$$ZA = 2.885 \times 10^{23} \times 2 \times 10^4$$

$$= 5.77 \times 10^{27} \text{ collisions sec}^{-1}$$

The area of the orifice = $A = \pi r^2 = 3.14 \times (0.2965/2)^2 = 0.0692$ cm^2.

$$\mu = \frac{\Delta m}{A \Delta t} = \frac{9.57 \times 10^{-3}}{0.0692 \times 110.5 \times 60} = 2.085 \times 10^{-5} \text{ g cm}^{-2} \text{ sec}^{-1}$$

$$P = \mu \sqrt{\frac{2\pi RT}{M}} = 2.09 \times 10^{-5} \sqrt{\frac{2 \times 3.14 \times 8.31 \times 10^7 \times 1555.4}{44.96}}$$

$$= 2.80 \text{ dynes cm}^{-2}$$

$$P = 2.80 \times 9.87 \times 10^{-7} = 2.77 \times 10^{-6} \text{ atm}$$

Solution 89. (*a*) $\frac{T_A}{T_B}$ (*b*) $\left(\frac{T_A M_B}{T_B M_A}\right)^{1/2}$

(c) $\frac{\sigma_B^2 C_B}{\sigma_A^2 C_A}$ (d) $\left(\frac{T_A M_B}{T_B M_A}\right)^{1/2} \frac{\sigma A^2 C_A}{\sigma B^2 C_B}$

Solution 90. (*a*) Gas viscosity results from the lateral transport of momenturm between layers moving at different speeds. The higher the temperature, the faster the molecules move and the more momentum they transport, per unit time.

(*b*) The larger the molecule, the more frequently it collides with other molecules, the shorter is its mean free path, and the less is the average momentum difference between the layers of gas between which the molecule travels in each free path. Therefore viscosity decreases as molecular diameter increases.

(*c*) As the pressure decreases, the number of momentum transporting molecules decreases proportionately. However, the mean free path varies inversely with the pressure and thus as the pressure decreases, the effectiveness of each molecule in transporting momentum between layers moving at different speeds increases. These two effects just compensate. At very low pressure, however the mean free path becomes comparable to the dimensions of the container. It can increase no further aand the viscosity approaches zero as the pressure (and thus the number of molecules) approaches zero.

Solution 91. (a) $\lambda = \dfrac{1}{\sqrt{2}\pi C\sigma^2}$

$$= \frac{1}{\sqrt{2}\,\pi\,(10^{-5}\text{atom cm}^{-2})\,(4\times 10^{-16}\ \text{cm}^2)} \times \frac{1\text{light–yr}}{9.464\times 10^{17}\text{cm}}$$

$$= 59 \text{ light-yr}$$

(*b*) $\bar{\upsilon} = \sqrt{\dfrac{8RT}{\pi M}}$

$$= \sqrt{\frac{8\times 8.314\times 10^7\times 7.8\times 10^5\text{erg (g atom)}^{-1}}{\pi\times 1.008g(\text{g } atom)^{-1}}}$$

$$= \sqrt{1.64\times 10^{14}\ \text{erg g}^{-1}} = 1.28\times 10^{7}\text{cm sec}^{-1}$$

$$t = \frac{\lambda}{\bar{\upsilon}} = \frac{59\text{ light–yr}\times 9.464\times 10^{17}\text{ cm(light–yr)}^{-}}{1.28\times 10^7\text{cm sec}^{-1}}$$

$$\times \frac{1\text{yr}}{3.156\times 10^7\text{sec}}$$

$$= 1.38\times 10^5$$

(Only approximate because the average of the ratio is not the ratio of the averages.)

Solution 92. (*a*) The Boyle temperature of a gas according to one definition is the temperature at which the so-called "second virial coefficient," B(T) or b(T), is zero. By examination of equation (a) we see that this condition is fulfilled when the quantity $[b - (A/RT^{3/2})] = 0$. This will happen when $T = T_B = (A/bR)^{2/3}$.

(*b*) For equation (*b*) it can be seen that the equation will reduce to $P\bar{V}^2/RT = 1$ (and thus Boyle's law will be obeyed) when the quantity $[1 - 6(T_c/T)1^2] = 0$. This will happen when $T = T_B = T_c\sqrt{6}$S.

Solution 93.

$$P = \frac{nRT}{V-nb} - \frac{n^2a}{V^2}$$

$$PV = nRT\left(\frac{V}{V-nb}\right) - \frac{n^2a}{V}$$

$$\left(\frac{\partial\,[PV]}{\partial P}\right)_T = \left[\frac{nRT}{V-nb} - \frac{nRTV}{(V-nb)^2} + \frac{n^2a}{V^2}\right]\left(\frac{\partial V}{\partial P}\right)_T$$

$(\partial V/\partial P)\ T < 0$ under all conditions

If $(\partial\ [PV]/\partial P)_T = 0$, as it should at $P = 0$, $T = T_B$, then the expression in square brackets equals 0, and hence

$$\frac{nRT}{V-nb} - \frac{nRTV}{(V-nb)^2} + \frac{n^2a}{V^2} = 0$$

$$nRTV^2(V-nb) - nRTV^3 + n^2a(V-nb)^2 = 0$$

$$-nRT[V^2(V-nb) - V^3] = n^2a(V-nb)^2$$

$$-RT(V^3 - nbV^3 - V^3) = na(V-nb)^2$$

$$RT = \frac{a}{b}\left[\frac{(V-nb)}{V}\right]^2 = \frac{a}{b}\left(1 - \frac{nb}{V}\right)^2$$

At the Boyle point $T = T_B$, $P = 0$, and $V = \infty$, so

$$RT_B = \frac{a}{b} \text{ and } T_B = \frac{a}{Rb}$$

An alternate method is to let $c = n/V$ and proceed as follows:

$$P = \frac{nRT}{V - nb} - \frac{n^2 a}{V^2}$$

$$PV = nRT\left(\frac{V}{V - nb}\right) - \frac{n^2 a}{V}$$

Since $c = n/V$, then

$$P = \frac{cRT}{1 - bc} - ac^2$$

and

$$PV = n\,\frac{RT}{1 - bc} - ac$$

$$\left(\frac{\partial\,[PV]}{\partial P}\right)_T = \left(\frac{\partial\,[PV]}{\partial c}\right)_T \left(\frac{\partial c}{\partial P}\right)_T = n\left[\frac{bRT}{(1 - bc)^2} - a\right]\left(\frac{\partial c}{\partial P}\right)_T$$

$$\left(\frac{\partial c}{\partial P}\right)_T = \frac{1}{(\partial P/\partial c)_T}$$

$$= \left[\frac{bcRT}{(1 - bc)^2} + \frac{RT}{1 - bc} - ac^2\right]^{-1} \underset{c \to 0}{P \to 0} \quad \frac{1}{RT} \neq 0$$

Therefore at the Boyle temperature the factor

$$bRT/(1 - bc)^2 - a,$$

evaluated at c = 0, must vanish.

$$bRT_B - a = 0, \; T_B = \frac{a}{bR}$$

Solution 94. Expanding the Dieterici equation by means of the infinite series for e^x , we obtain

$$P = \frac{RT}{(\overline{V} - b)} - \frac{a}{\overline{V}(\overline{V} - b)} + \frac{a^2}{2RT\,\overline{V}^2(\overline{V} - b)} - \ldots$$

At low densities it is reasonable to assume that the third and subsequent terms are neglible and that $\overline{V}(\overline{V}-b) \approx \overline{V}^2$. Thus we obtain

$$P = \frac{RT}{(\overline{V}-b)} - \frac{a}{\overline{V}^2}$$

which is identical to the van der Waals equation.

Solution. 95

$$P = \frac{RT}{\overline{V}-bT} - \frac{a}{\overline{V}^3}$$

$$\left(\frac{\partial P}{\partial \overline{V}}\right)_T = \frac{-RT}{(\overline{V}-bT)^2} + \frac{3a}{\overline{V}^4} = 0$$

$$\left(\frac{\partial^2 P}{\partial \overline{V}^2}\right)_T = \frac{2Rt}{(\overline{V}-bT)^3} - \frac{12a}{\overline{V}^5} =$$

$$\frac{RT_c}{(\overline{V}_c - bT_c)^2} = \frac{3a}{\overline{V}_c^4}$$

$$\frac{2RT_c}{(\overline{V}-bT_c)^3} = \frac{12a}{\overline{V}_c^5}$$

Dividing

$$\frac{2}{\overline{V}_c - bT_c} = \frac{4}{\overline{V}_c}$$

$$b = \frac{\overline{V}_c}{2T_c}, \quad a = \frac{RT_c\overline{V}_c^4}{3(\overline{V}_c - bT_c)^2} = \frac{4RT_c\overline{V}_c^{\,2}}{3}$$

Solution 96.

$$P\overline{V} = RT + APT - BP$$

$$P = \frac{RT}{\overline{V} - AT + B}$$

(*a*) $$P\overline{V} = \frac{RT}{1 + [(B - AT)/\overline{V}]}$$

$$\approx RT\left[1 - \frac{B - AT}{\overline{V}}\right] \text{ if } \left|\frac{B - AT}{\overline{V}}\right| << 1$$

At the Boyle temperature T_B

$$\lim_{V \to \infty}\left[\frac{\partial (P\overline{V})}{\partial (1/\overline{V})}\right] = 0$$

$$B - AT_B = 0$$

$$T_B = \frac{B}{A}$$

(b) $$\left(\frac{\partial P}{\partial \overline{V}^2}\right)_T = \frac{-RT}{(\overline{V} - AT + B)^2}$$

$$\left(\frac{\partial^2 P}{\partial \overline{V}^2}\right)_T = \frac{2RT}{(\overline{V} - AT + B)^3}$$

At the critical point $(\partial P/\partial \overline{V})_T = 0$ and $(\partial^2 P/\partial \overline{V}^2)_T = 0$. For this gas these derivatives can be 0 only when T = 0 or $\overline{V} - AT + B$ is infinite. Neither case corresponds to a real critical point.

Solution 97. At the critical point, $(\partial P/\partial \overline{V})_T = 0$ and $(\partial^2 P/\partial \overline{V}^2)_T = 0$.

$$\left(\frac{\partial P}{\partial \overline{V}}\right)_T = \frac{-RT}{(\overline{V} - b)^2} + \frac{a}{\overline{V}^2} = 0$$

or

$$\frac{RT}{(\overline{V} - b)^2} = \frac{a}{\overline{V}^2}$$

$$\left(\frac{\partial^2 P}{\partial \overline{V}^2}\right)_T = \frac{2RT}{(\overline{V} - b)^3} - \frac{2a}{\overline{V}^3} = 0$$

or $$\frac{2RT}{(\overline{V} - b)^3} = \frac{2a}{\overline{V}^3}$$

Divide the second equation by the first: $2/(\bar{V}-b) = 2/\bar{V}$, from which b = 0. Therefore the two derivatives can never vanish simultaneously unless b = 0, contrary to the stipulation that *a* and *b* are distinct from zero, and the gas has no critical point.

Solution 98. $P = \dfrac{RT}{\phi V_c - b}\, e^{-n/(R\phi V_c T_c)}$

$$\pi \frac{a}{4e^2b^2} = \frac{R\theta\,(a/4Rb)}{\phi\,(2b) - b} \exp\left(\frac{-a}{R\phi 2b\theta a/4Rb}\right)$$

$$\pi = \frac{\theta}{2\phi - 1}\, e^{2 - 2/\phi\,\theta}$$

Solution 99. Choose as the volume unit a volume equal to 1 mm times the cross-sectional area of this barometer tube. In these units the volume is numerically equal to the tube length in millimeters. The pressure will be in torr. The correction ΔP to be applied to the reading is equal to the pressure exerted by the entrapped air: $\Delta P = KT/V$, where = $V = 780 - P_{\text{apparent}}$, and K is a constant as yet unknown. To determine K, we apply this equation to the conditions given :

$$K = \frac{V\Delta P}{T} = \frac{(780 - 717)\times 3.0}{293.2} = 0.64$$

Then

$$\Delta = \frac{0.64T}{780 - P_{\text{apparent}}}$$

Solution 100. Let P_2 = the final pressure of the entrapped air, in cm Hg

h = the final height of the Hg in the cylinder, in cm

Then

$$P_2 = 76\times \frac{100}{100 - h}$$

Also

$$P2 = 100 - h + 76$$

$$\frac{76 \times 100}{100 - h} = 176 - h$$

$$h^2 - 276h + 10,000 = 0$$

(*a*) h = 42.9 cm, 233 cm (the 233 root is extraneous)

(*b*) $P_2 = 176 - 42.9 = 133$ cm Hg

Solution 101. Let P_2 = the final pressure of the entrapped air, in cm Hg

h = the final height of the Hg in the long arm, in cm

Then

$$P_2 = P_1 \frac{V_1}{V_2}$$

$$P_2 = 76 \times \frac{100}{100 - h}$$

$$P_2 = 50 - h + 76$$

$$\frac{76 \times 100}{100 - h} = 50 - h + 76$$

$$h_2 - 226h + 5000 = 0$$

h = 25.4 cm, 201 cm (201 is extraneous).

(*b*) $P^2 = 76 \times \frac{100}{100 - 25.4} = 102$ cm Hg

Solution 102. The total number of moles of N_2 in the whole system is 2.00. Let n = the final number of moles of N_2 in the 100°C sphere (Sphere 1). Then $2 - n$ = the final number of moles of N_2 in the 0°C spherre (Sphere 2).

$$V_1 = V_2 = 22.4 \text{ liters}$$

$$T_1 = 373^\circ K$$

$$T_2 = 273^\circ K$$

$$P_1 = P_2$$

$$\frac{nRT_1}{V_1} = \frac{(2-n)RT_2}{V_2}$$

$$(2 - n)273 = 373n$$

n = 0.846 moles = number of moles in sphere 1

$$P_1 = P_2 = \frac{0.846 \times 0.08206 \times 373.2}{22.4} = 1.16 \text{ atm}$$

$$2 - n = 2 - 0.846 = 1.154 \text{ moles}$$

= number of moles in sphere 2

Solution 103. The error is in step 5, where equations 3 and 4 are combined. Equation 3 assumes constant temperature but not constant volume or constant pressure. Equation 4 assumes constant pressure but not constant volume or constant temperature. Equations 3 and 4 should be (3′) $V = [k_1(T)]/P$ and (4′) $V = k_2(P)$ T, where k_1 and k_2 are unspecified functions of T and P, respectively. Now when the equations are combined, we obtain (5′)

$$\frac{k_1(\text{T})}{\text{P}} = k_2\,(\text{P})\text{T} \quad \text{or} \quad \frac{k_1\,(\text{T})}{\text{T}} = k_2\,(\text{P})\,\text{P}$$

The left side does not depend on P and the right side does not depend on T. But the two sides are equal for all T and P; therefore neither side depends on T or P —both are constants:

$$\frac{k_1(T)}{T}\ K_1,\ \ k_1(T) = K_1 T$$

$$k_2(P)P = k_2,\ \ k_2(P) = \frac{K_2}{P}$$

We have now ascertained the forms of the functions k_1 and k_2. When these are substituted into (5′), we obtain $K_1 = K_2$. Then from (3′) or (4′) we obtain $V = (K_1T)/P$.

Solution 104. Plot density of mixture in g liter^{-1} vs composition of mixture in mole per cent. This will be a linear plot with slope = $[P(M_1 - M_2)]/100$ RT; the intercept on the density axis when

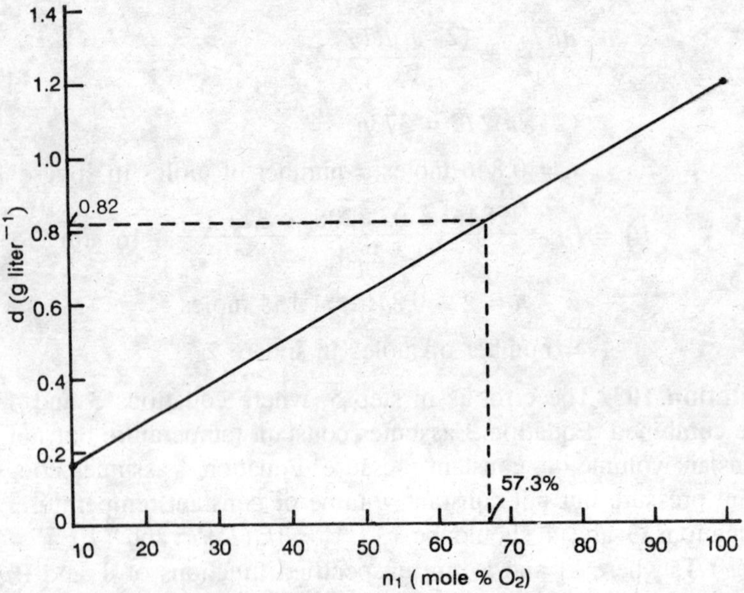

Fig.

$n_1 = 0$ will be PM_2/RT, and the intercept on the density axis when $n_2 = 0$ will be PM_1/RT. The reasoning is as follows:

Let

M = the apparent molecular weight of the mixture

M_1 = the molecular weight of oxygen

M_2 = the molecular weight of helium

n_1 = the mole pe rcent of oxygen in the mixture

n_2 = the mole per cent of helium in the mixture

W = the weight of the mixture

d = the density of the mixture

Then

$$PV = nRT = \frac{WRT}{M}$$

$$d = \frac{W}{V} = \frac{PM}{RT}$$

$$M = \frac{M_1 n_1}{100} + \frac{M_2 n_2}{100}$$

$$n_2 = 100 - n_1$$

$$M = \frac{M_1 n_1 + M_2 (100 - n_1)}{100} = \frac{n_1 (M_1 - M_2) + 100M_2}{100}$$

$$d = \frac{P}{RT}\frac{n_1(M_1 - M_2) + 100M_2}{100} = \frac{P(M_1 - M_2)n_1}{100RT} + \frac{PM_2}{RT}$$

The last equation is the algebraic representation of the desired plot. It can be seen by inspection that this equation gives the slope and $n_1 = 0$ intercept specified above. If we had eliminated n_1 instead of n_2, we would have an analogous equation showing that the density intercept at $n_2 = 0$ is PM_1/RT, as specfied above.

(*b*) See the figure. below At $n_1 = 100$

$$d = \frac{PM_2}{RT} = \frac{1\times 32.0}{0.0821\times 298} = 1.31 \text{ g liter}^{-1}$$

At $n_1 = 0$

$$d = \frac{PM_1}{RT} = \frac{1\times 4.00}{0.0821\times 298} = 0.163 \text{ g liter}^{-1}$$

We draw a straight line between these two points and read off a composition of 57.3 mole percent O_2 corresponding to a density of 0.820 g liter^{-1}

Solution 105.

(*a*) $$\overline{\upsilon_x^4} = \frac{\int_{\infty}^{\infty} \upsilon_x^4 \exp(-\frac{1}{2} m\upsilon_x^2/kT)\, d\upsilon_x}{\int_{-\infty}^{\infty} \exp(-\frac{1}{2} m\upsilon_x^2/kT) d\upsilon_x}$$

(*b*) $$\overline{\upsilon^4} = \frac{\int_0^{\infty} \upsilon^4 e^{-m\upsilon^2/2kT} \upsilon^2\, d\upsilon}{\int_0^{\infty} e^{-m\upsilon^2/2kT} \upsilon^2\, d\upsilon}$$

(*c*) The change in limits of intergration is inessential; if the limits in (a) were 0 to ∞, the result would be the same. The important difference is the presence of an additional factor υ^2 in the integrand of (b). This factor appears because the molecules with speeds between υ and $\upsilon + d\upsilon$ include all those with representative points (in velocity space) in a spherical shell of volume $4\pi\,\upsilon^2\,d\upsilon$. Equivalently the integrals in (*b*) are evaluated in spherical coordinates, in which the volume elements is $\upsilon^2 \sin\theta \sin\, d\upsilon\, d\theta\, d\varphi$; the angular integrals are the same in numerator and denominator.

Solution 106. $P\overline{V} = RT + bP$; slope $= [\partial(P\overline{V})/\partial P]_T = b$ (see figure).

Significance of slope : slope = the constant b in the equation of state.

Significance of intercept: at P = 0, $P\overline{V} = RT$ (ideal gas behavior at zero pressure). For a real gas if $b > 0$, the gas would have to be above its Boyle temperature if this equation of state were to apply at low pressures inasmuch as the slope is positive near $P = 0$.

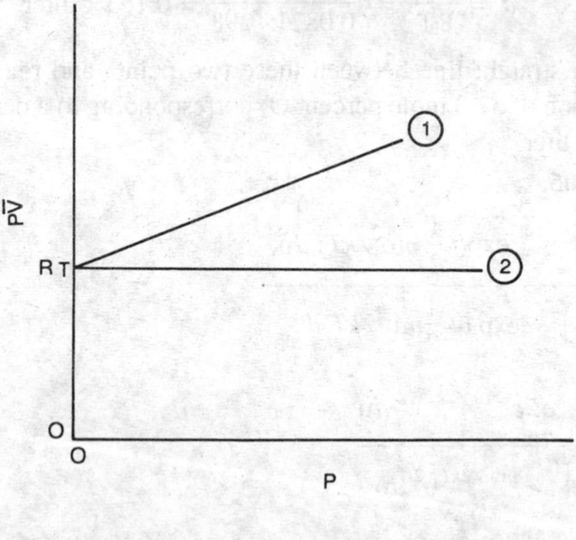

Fig.

Solution 107. $dV = \left(\frac{\partial V}{\partial P}\right)_P dP + \left(\frac{\partial V}{\partial T}\right)_P dT$

From α

$$\left(\frac{\partial V}{\partial T}\right)_T = \frac{k_1 V C_P T^{(C_P/C_V)-1}}{C_V}$$

and from β

$$\left(\frac{\partial V}{\partial P}\right)_T = \frac{-k_2 V}{P}$$

$$\frac{dV}{V} = -k_2 \frac{dP}{P} + k_1 \frac{C_P}{C_V} T^{(C_P/C_V)-1} dT$$

$$\ln V = -K_2 \ln P + k_1 \frac{C_P}{C_V} \cdot \frac{C_V}{C_P} T^{C_P/C_V} + \text{constant}$$

$$= -k_1 \ln P + k_1 T^{C_P/C_V} + \text{constant}$$

Therefore the desired equation of state is

$$P^{k_2} V = k \exp(kT^{C_P/C_V})$$

Solution 108. We shall present two different methods for doing this problem. The first method depends on the expansion of the van der Waals equation into cubic form at the critical point and comparing coefficients of V with the equation $(V - V_c)^3 = 0$. The details are as follows:

(1) The van der Waals equation

$$\left(\frac{P - n^2 a}{V^2}\right)(V - nb) = nRT$$

(2) Expanding (1)

$$PV - \frac{n^2 a}{a} - Pnb - \frac{n^3 ab}{V^2} = nRT$$

(3) Multiplying (2) by V^2

$$PV^3 + n^2 av - PnbV^2 - n^3 ab = nRTV^2$$

(4) Dividing (3) by P

$$V^3 + \frac{n^2aV}{P} - nbV^2 - \frac{n^3ab}{P} = \frac{nRTV^2}{P}$$

(5) Rearranging (4) in descending powers of V,

$$V^3 - \left(\frac{nRT}{P} + nb\right)V^2 + \frac{n^2aV}{P} - \frac{n^3ab}{P} = 0$$

(6) Now at the critical point all three V roots are equal ; hence $V_1 = V_2 = V_3 = V_c$, as for the 3 roots of $(V - Vc)^3 = 0$

(7) Expanding (6)

$$(V - V_c)^3 = V^3 - 3V_cV^2 + 3V_c^2V - V_c^3 = 0$$

(8) At the critical point

$$V^3 - \left(\frac{nRT_c}{P} + nb\right)V^2 + \frac{n^2aV}{P_c} - \frac{n^3ab}{P_c} = 0$$

Now, comparing coefficients of V^3, V^2, V^1 and V^0 in (7) and (8), we obtain

(9) $V^3 : 1 = 1$

(10) $V^2 : 3V_c : = nRT_c/P_c + nb$

(11)$V^1 : 3V_c^2 = n^2a/P_c$

(12) $V^0 : V_c^3 = n^3ab/P_c$

(13) Equating P_c's in (11) and (12) we obtain $n^2a/3V_c^2 = n^3\,ab/V_c^3$, which readily reduces to $V_c = 3nb$.

(14) Substituting (13) into (11)

$$P_c = \frac{n^2a}{3V_c^2} = \frac{n^2a}{27n^2b^2} = \frac{a}{27b^2}$$

Alternatively we could have substituted (13) into (12) to get the same final result.

(15) By substituting (13) and (14) into (10)

$$3V_c = \frac{nRT_c}{P_c} + nb$$

$$T_c = \frac{P_c(3V_c - nb)}{nR} = \frac{a(9nb - nb)}{27b^2nR} = \frac{8a}{27Rb}$$

The second method depends upon the generation of two new equations by means of partial differentiation.

(1′) First, we start with the van der Waals equation in the form

$$P = \frac{nRT}{(V - nb)} - \frac{n^2a}{V^2}$$

(2′) Then we take the first derivative to obtain

$$\left(\frac{\partial P}{\partial V}\right)_T = \frac{-nRT}{V - nb)^2} + \frac{2n^{2a}}{V^3}$$

(3′) And we differentiate (2′) to obtain the second derivative

$$\left(\frac{\partial^2 P}{\partial P^2}\right)_T = \frac{2nRT}{(V - nb)^3} - \frac{6n^2a}{V^4}$$

At the critical point the isothermal curve that represents P as a function of V has a horizontal point of inflection. Because the curve is horizontal, $(\partial P/\partial V)_T = 0$; and because it has a point of inflection, $(\partial^2 P/\partial^2)_T = 0$. Then

$$(4') \quad \frac{RT}{(V - nb)^2} = \frac{2na}{V^3}$$

$$(5') \quad \frac{2RT}{(V - nb)^3} = \frac{6na}{V^4}$$

Dividing the left side of equation (4′) by the right side of (5′), and similarly for the right sides, we have

$$(6'), \quad \frac{(V - nb)}{2} = \frac{V}{3} \text{ and } V_c = 3nb$$

Substituting this result into (4′), we obtain

$$(7')\ \frac{RT}{(2nb)^2} = \frac{2na}{(3b)^3} \text{ and } T_c = \frac{8a}{27Rb}$$

(8′) And, finally, from (1′)

$$P_c = \frac{8a}{27(b)(2b)} - \frac{a}{(3b)^2} = \frac{a}{27b^2}$$

The advantage of the first method is that it does not require the use of the calculus. The advantage of the second method is that it is simpler, requires less mathematical intuition, and is more generally applicable to other problems of a smilar nature.

Solution 109. At the critical point (if any) $(\partial P/\partial V)_T = 0$ and $(\partial^2 P/\partial V^2)_T = 0$. The equation can be solved (as a quadratic equation) for P but it is simpler to obtain the first derivative by the formula

$$\left(\frac{\partial P}{\partial V}\right)_T = -\frac{(\partial T/\partial V_P}{(\partial T/\partial P)_V} = \frac{-(P + aP^2)}{(V-b)(1+2aP)}$$

Before we go to the trouble of evaluating $(\partial P/\partial V^2)_T$, we should consider whether, or in what cases, the first derivative vanishes. $(\partial P/\partial V)_T = 0$ if, and only if, $P + aP^2 = 0$. But $P + ap^2 = 0$ only when $T = 0$, and the gas does not have a critical point.

Solution 110. (1) $P = \dfrac{RT}{(V-b)} - \dfrac{a}{T\overline{V}^2}$

$$(2)\ \left(\frac{\partial P}{\partial \overline{V}}\right)_T = \frac{-RT}{(\overline{V}-b)^2} + \frac{2a}{T\overline{V}^3} = 0$$

$$(3)\ \left(\frac{\partial^2 P}{\partial \overline{V}^2}\right)_T = \frac{2RT}{(\overline{V}-b)^3} - \frac{6a}{T\overline{V}^4} = 0$$

From (2) and (3)

$$\frac{2}{\overline{V}_c - b} = \frac{3}{\overline{V}_c} \text{ and } b = \overline{V}_c/3$$

From (2) we get

$$(4)\quad a = \frac{RT_c^2\,\overline{V}_c^3}{2(\overline{V}_c - b)^2} = \frac{9RT_c^2\overline{V}_c}{8}$$

Substituting a and b into (1) and simplifying

$$P_c = \frac{3RT_c}{8\overline{V}_c};\quad R = \frac{8P_c\overline{V}_c}{3T_c}$$

Substituting R into (4)

$$a = 3P_c\overline{V}_c^2T_c$$

Then (1) becomes, in terms of reduced variables (π, θ,Φ) and critical constants,

$$\pi P_c = \frac{(8P_c\overline{V}_c/3T_c)\,(\theta T_c)}{\phi\overline{V}_c - \frac{1}{3}\overline{V}_c} - \frac{3P_c\overline{V}_c^2T_c}{\theta T_c\,\phi^2\,\overline{V}_c^2}$$

$$\pi = \frac{8\theta}{3\phi - 1} - \frac{3}{\pi\phi^2}$$

Solution 111. We shall proceed as in Problem 1-III-10, method 2, by taking first and second derivatives to generate two new equations.

$$(1)\quad P = \frac{RT}{(\overline{V} - b)}\,e^{-A/RT^{3/2}p}$$

$$(2)\quad \left(\frac{\partial P}{\partial \overline{V}}\right)_T = \frac{RT}{(\overline{V} - b)}\,e^{-A/RT^{3/2}P}\left(\frac{A}{RT^{3/2}\overline{V}^2} - \frac{1}{(\overline{V} - b)}\right)$$

$$(3)\quad \left(\frac{\partial^2 P}{\partial \overline{V}^2}\right)_T = \left\{\left[\frac{A}{RT^{3/2}\overline{V}^2} - \frac{1}{(\overline{V} - b)}\right]\left[\frac{-1}{(\overline{V}} - b)^2 + \frac{A}{RT^{3/2}\,\overline{V}^2}\right]\right.$$

$$\left. + \left(\frac{1}{(\overline{V} - b)}\right)\left[\frac{-2A}{RT^{3/2}\overline{V}^3} + \frac{1}{(\overline{V} - b)^2}\right]\right\}$$

$$\times\, RT^{e^{-A/RT^{3/2}\overline{V}}}$$

At the critical point, $(\partial P/\partial V)_T = 0$ and $(\partial^2 P/\partial V)_T = 0$. In (2) there is only one factor that can vanish.

$$(4)\quad \frac{A}{RT_c^{3/2}\overline{V}_c^3} - \frac{1}{\overline{V}_c - b} = 0$$

The first term in (3) is then 0 also. For the second term to vanish we must have

$$(5)\quad \frac{-2A}{RT_c^{3/2}\overline{V}_c^3} + \frac{1}{(\overline{V}_c - b)^2} = 0$$

From (4) and (5) $2/\overline{V}_c = 1/(\overline{V}c - b)$, $\overline{V}_c = 2b$. Substituting into (4), $T_c^{3/2} = A/4bR$, or $T_c = (A/4bR)^{2/3}$. Finally, substituting $\overline{V}_c$ and T_c into (1),

$$P_c\ \frac{R}{b}\left(\frac{A}{4bR^{2/3}}\right) e^{-2}$$

Solution 112. (*a*) $\dfrac{P\overline{V}}{RT} = \dfrac{1}{1 - B/\overline{V}} - \dfrac{A\overline{V}^{m+1}}{RT}$

(*b*) $\overline{V}^{m+1}$ must approach 0 as $\overline{V}$ becomes infinite. This requires that $m + 1 < 0$, or $m < -1$.

(*c*)

$$(1)\quad P = \frac{RT}{\overline{V} - B} - A\overline{V}^m$$

$$(2)\quad \left(\frac{\partial P}{\partial \overline{V}}\right)_T = \frac{-RT}{(\overline{V} - B)^2} - mA\overline{V}^{m-1}$$

$$(3)\quad \left(\frac{\partial^2 P}{\partial \overline{V}^2}\right)_T = \frac{2RT}{(\overline{V} - B)^3} - m(m-1)A\overline{V}^{m-2}$$

(*d*) At the critical point $(\partial P/\partial \overline{V})_T = 0$ and $(\partial P/\partial \overline{V}^2)_T = 0$. We write (2) and (3), then, as

$$(2')\quad \frac{-2RT}{(\overline{V} - B)^2} = mA\overline{V}^{m-1}$$

$$(3')\quad \frac{2RT}{(\overline{V} - B)^3} = m(m-1)A\overline{V}^{m-}$$

and divide (3′) by (2′)

$$\frac{-2}{(\overline{V}-B)} = \frac{m-1}{\overline{V}}$$

$$(4)\ \overline{V}_c = \left(\frac{m-1}{m+1}\right) B$$

(e) If we already know that $m < -1$, then we also know that $m - 1$ and m + 1 are both negative, and $\overline{V}_c$ has the same sign as B. In order for $\overline{V}_c$ to be positive, B must be positive. (If we do not take account of the result of (b), all we can say is that $(m - 1)/(m + 1)$ and B must have the same sign. If $m < -1$ or $m > +1$, $B < 0$; if $-1 < m < +1$, $B < 0$. If m = –1, $\overline{V}_c$ is infinite; if m = + 1, $\overline{V}_c = 0$, regardless of B.)

(*f*) From (2′)

$$T_c = \frac{-(\overline{V}_c - B)^2\, mA\overline{V}_c^{m-1}}{R}$$

From (4)

$$V_c - B\ \frac{-2}{m+1}\ B$$

Then

$$9T_c = \frac{-4B^2 mA\overline{V}^{m-1}}{(m+1)^2 R} = \frac{-4B^{m+1}m(m-1)^{m-1}A}{(m+1)^{m+1}R}$$

(*g*) Since $m < 0$ and B > 0, Tc will be > 0 if, and only if, A > 0. This is more easily seen in the first form of the result than in the second.

(*h*) B is related to the volume occupied by the molecules. The requirement B > 0 means merely that the molecules must have a positive volume. The term containing A and m is intended to correct for intermolecular forces. If A < 0, then P is increased by these forces, which must in this case be repulsive. A gas in which the molecules repel would never condense and there would be no critical point; accordingly we find a real critical point only when A > 0 corresponding to attraction and a diminution in pressure by the term $A\overline{V}^{n}$. In order for the gas to behave ideally at low pressure the attractive forces must decrease, and decrease rapidly enough, as the distance between molecules increases, and thus as $\overline{V}$ increases. The condition m < –1 results from this requirements.

Part III
PROBLEM STATEMENTS

CHEMICAL KINETICS

Problem 1 : The hydrogenation of ethylene

$C_2H_4 + H_2 \rightarrow C_2H_6$

in the presence of mercury vapour is thought to proceed through the following steps.

$$Hg + H_2 \xrightarrow{k_1} Hg + 2H \quad (1)$$

$$H + C_2H_4 \xrightarrow{k_2} C_2H_5 \quad (2)$$

$$C_2H_5 + H_2 \xrightarrow{k_3} C_2H_5 + H \quad (3)$$

$$H + H \xrightarrow{k_4} H_2 \quad (4)$$

Assuming that H and C_2H_5 attain equilibrium determine the rate of formation of C_2H_6 in terms of the rate constants and concentrations (H*g*), (H_2) and (C_2H_4)

Problem 2 : Determine *a* and *b* in the rate equation

$$\text{rate} = kC_A^a C_B^b$$

given the following :

Table 1 :

rate, M s^{-1}	0.05	0.10	0.20	0.40
$C_{A,0}$, M	1	1	2	2
$C_{B,0}$, M	1	2	1	2

Calculate k for this reaction.

Problem 3 : For the chemical reaction

$$H_2O + C_rO_7{}^{2-}\,(aq) \rightarrow CrO_4{}^{2-}(aq) + 2H^+\,(aq)$$

evaluate the various expressions for the rate in terms of $-dc_{cr_2O_7^{2-}}$ /dt. If the actual reaction equation for the chromate dichromate system is identical to the stoichiometric equation, write the rate equation. If $C_{H_2O}O$ is larage enough to remain essentilly constant, write the revised rate equation. What is the overall order of both of these rate equation?

Problem 4 : A proposed mechanism for the reaction between H_2 *g) and* B_{r2} *(g)* $k1$ $2Br$ is

$$Br_2 \underset{k_5}{\overset{k_1}{\rightleftarrows}} 2Br$$

$$Br + H_2 \xrightarrow{k_2} HBr + H$$

$$H + Br_2 \xrightarrow{k_3} HBr + Br$$

$$H + HBr \xrightarrow{k_4} H_2 + Br$$

Write expressions for dc_{HBr}/dt, dc_H/dt and dc_{Br}/dt. Assuming that dcH/dt = $dcBr/dt$ = O solve for d CHBr/ in terms of C_{H_2} , C_{Br_2} and C_{HBr}.

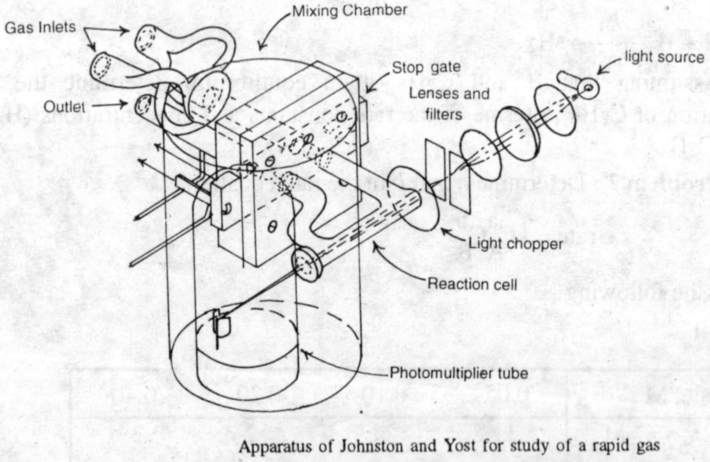

Apparatus of Johnston and Yost for study of a rapid gas reaction : isometric projection of mixing chamber, stop gate, and 2 mm diameter reaction cell, with schematic drawing of the light source, filters, and lenses.

Problem 5 : For the series of competing reactions $H + HO_2 \xrightarrow{k_1} H_2 + O_2$; $H_2 + HO_2 \xrightarrow{k_2} 2OH$; $H + HO_2 \xrightarrow{k_3} H_2o + O$ Westenberg and deHass report $k_1/k_2/k_3 = 0.62/0.27/0.11$. find the ratio of the products time t.

Problem 6 : Assuming that $dC_c/dt = 0$ for the reactions

$$A + B \underset{k_{-1}}{\overset{k_1}{\rightleftarrows}} C$$

$$C + B \xrightarrow{k_2} D$$

find $-dC_A/dt$, $-dC_B/dt$ and dC_D/dt in terms of C_A, C_B and C_D.

Problem 7 : For the reaction

$$2NO + O_2 \rightarrow 2NO_2$$

the following reaction mechanism have been suggested;

1) $NO + NO \xrightarrow{k_1} N_2O_2$

2) $N_2O_2 \xrightarrow{k_2} 2NO$

3) $N_2O_2 + O_2 \xrightarrow{k_3} 2NO_2$

a) Apply the steady state approximation to $[N_2O_2]$ and show that the rate law in given by

$$= \frac{2k_1k_3[NO]^2[O_2]}{k_2 + k_3[O_2]}$$

b) Calculate the overall activation energy for this reaction assuming that only a small fraction of the N_2O_2 formed in step 1) actually goes to form products in step 3) and most of the N_2O_2 reverts to NO in step 2). The activation energies for step 1) through 3) are 82*k*J and 205 *k*J and 82 *k*J respectively.

Problem 8 : The value of the activation energy, E_a, for the thermal decomposition of gaseous acetaldehyde is 45, 500 cal mole^{-1} and the molecular diameter, σ, of a molecule of acetaldehyde is 5×10^{-8} cm. This decomposition is a second order reaction.

(a) Determine the number of molecules colliding per milliliter per second at 800 and 760 torr pressure

(b) Calculate the second order rate constant, k, in liters mole^{-1} sec^{-1}.

Problem 9 : The half-life for a given reaction was halved as the initial concentration of a reactant was doubled. What is n for this component ?

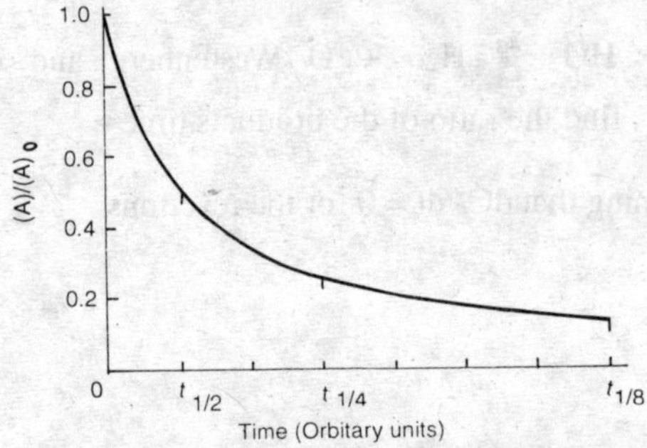

The second-order rate law, (A)/(A)o as a function of time, showing the successing doubling of $t_{s/s}$

Promblem 10 : At 393.7°C, the second-rate constant for the decomposition of gaseous hydrogen iodide in a second order reaction is $2.6 \times 10^{-4} M^{-1} s^{-1}$. Given that the Arrhenius activation energy for the reaction is 45.6 $k$cal mol^{-1}, calculate the second order rate constant for this reaction from collision theory. The molecular weight of HI = 127.9 g mol^{-1}, the collision diameter = 3.5 Å and the orientation factor, P = 1.

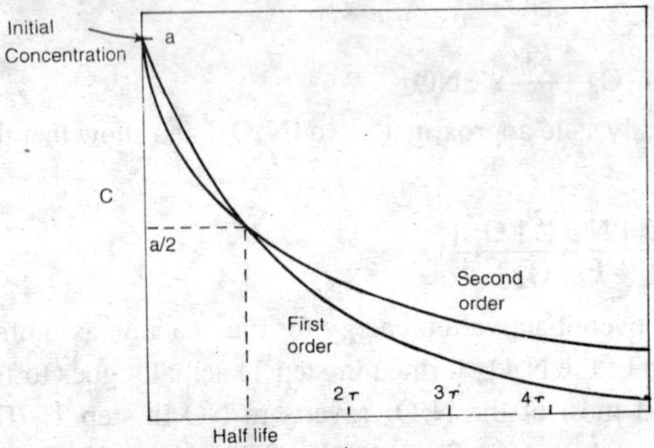

Concentration vs time curves for first- and second-order reactions with same initial concentrations ***a*** **and same half-lives τ**

Problem 11 : Measurement of the second order decomposition of acetaldehyde over the temperature range 700-850°K was made and the second order rate constant was given as

T (°K)	700	730	760	790	810	840	910	1000
k_2	0.011	0.035	0.105	0.343	0.789	2.17	20.0	145

Determine by a plot the activation energy and the pre-exponential factor for the decompostion.

Problem 12 : Consider the opposing second-order reactions

$$A_2 + B_2 \underset{k_{-2}}{\overset{k_2}{\rightleftarrows}} 2AB$$

Derive an expression for k in terms of k_2 and k_{-2} .

Problem 13 : At a temperature of 300°K it takes 12.6 min for a certain reaction to be 20% complete, and at 340°K the same reaction takes 3.20 min. Calculate the activation energy E_a for the reaction.

Problem 14 : The partial pressure of azomethane, $CH_3N_2CH_3$ was measured as a function of time at 600°*k* as follows

t(sec)	0	1000	2000	3000	4000
$P_{CH_3N_2CH_3}$ (*mm*Hg)	8.2×10^{-2}	5.7×10^{-2}	3.99×10^{-2}	2.78×10^{-2}	1.94×10^{-2}

a) Show that the decomposition, $CH_3N_2CH_3 \rightarrow CH_3CH_3 + N_2$, is a first-order reaction in azomethane.

b) Determine the rate coefficient at 600°K.

Problem 15 : Two equimolar solutions A and B with equal volume of solutions are mised together and the reaction A + B = C takes place. After 1 hr, A is 75% reacted. Calculate how much of A will not react after 2 hrs if a) the reaction is first order in A and zero order in B. b) the reaction is first order in both A and B and c) the reaction is zero order in both A and B.

Problem 16 : Chloroform reacts according to the following equation in the presence of sodium methoxide in methanol solution.

$$CHCl_3 + OMe^- \rightleftarrows CCl_3^- + MeOH \text{ (fast)}$$

$$CCl_3^- \rightarrow CCl_2 + Cl^- \text{ (slow)}$$

and $CCl_2 \rightarrow$ products (fast)

An experiment was done to measure the order of the slow step with respect to CCl_3^- and to determine the amount of Cl^- produced in time, t at 59.7°C by titration with 0.0100 N $AgNO_3$, V *ml* being required. The following results were obtained.

t(min)	V, *ml*
0	1.71

4	3.03
9	4.49
15	5.97
22	7.39
30	8.87
41	10.48
50	11.7
∞	15.98

Show that the reaction is first order and determine the rate constant.

Problem 17 : Consider the gaseous decomposition reaction of cyclopentene to H_2 and cyclopentadiene :

$$c\text{-}C_5H_8 = H_2 + c\text{-}C_5H_6$$

(a) How is dP/dt related to $-dC_{C_5H_8}/dt$? (b) If the reaction is first-order, what are the units on k ? (c) Derive the first-order integrated rate equation in terms of $P_{C_5H_8}$, 0 and P.

Problem 18 : In a reaction between sodium thiosulfate and *n*-propyl bromide at 37°C, the amount of unreacted thiosulfate was measured by titration with I_2. Data for the I_2 titer in cm^3 of 0.02572 N iodine per 10.02 cm^3 sample of reaction mixture is given below.

t(sec)	0	1110	2010	3192	5052	7380	11232	78840
I_2 (titer)	37.63	35.20	33.63	31.90	29.80	28.04	26.01	22.24

Write a balanced equation for the reaction; show that the reaction is second order and determine the rate constant.

Problem 19 : For the set of reactions

$$A + B \underset{k_{-1}}{\overset{k_1}{\rightleftarrows}} C \qquad C + B \xrightarrow{k_2} D$$

find $-dC_A/dt$, $-dC_B/dt$, dC_C/dt and dC_D/dt.

Problem 20 : Consider the following mechanism describing enzyme catalysis :

$$E + S \underset{k_{-1}}{\overset{k_1}{\rightleftarrows}} X \underset{k_{-2}}{\overset{k_2}{\rightleftarrows}} E + P$$

where E is the enzymatic site, S is the substrate, X is the enzyme substrate complex and P is the product of the reaction. Derive the rate equation for this process assuming $dC_X/dt = 0$ and discuss the results for the reaction during the initial stages.

Problem 21 : The Gibbs adsorption coefficent ψ for an adsorbate on the surface of an adsorbent is related to the surface tension, r, and the activity of the adsorbate, a_2, by the expression

$$\psi = -\frac{a_2}{RT}\frac{dr}{da_2} = -\frac{1}{RT}\frac{dr}{d(\ln a_2)}$$

If glycerol raises the surface tension of water as it is added, will the former be positively or negatively adsorbed?

Problem 22 : The luminescent reaction of luciferin catalyzed by the enzyme luciferase obtained from the crustacean Cypridina was measured at an enzyme concentration of 10^{-9} M. The initial luciferin concentration is given as cm^3 of 8.0×10^{-5}M luciferin in 20 cm^3 of reaction mixture.

Luciferin conc.	0.04	0.06	0.08	0.10	0.20	0.40	0.90
Initial rate (15°)	15	19	23	26	44	56	76
Millivolts per minute							
(22°C)	18	23	33	37	62	91	123

The conversion factor from millivolt reading of the photoelectric integrator to moles of luciferin reacted was 8.6×10^{-12} mol. mv^{-1} at 22°C and 7.7×10^{-12} at 15°C. For this reaction, estimate ΔH, ΔS and ΔH°, ΔS° for the formation of the enzyme substrate complex. What standard state is used?

Problem 23 : Williams and Petrucci studied the system

$$Ni(NCS)^+ + NCS^- \underset{k_r}{\overset{k_f}{\rightleftarrows}} Ni(NCS)_2$$

at 25°C in methanol using a pressure-jump technique and obtained the following data :

$C_{Ni(NCS)_2}$, M	0.001	0.002	0.005	0.010	0.025	0.05	0.10
τ, *ms*	4.08	3.74	2.63	1.84	1.31	0.88	0.67

Determine k_f, k_r and K. τ is the relaxation time.

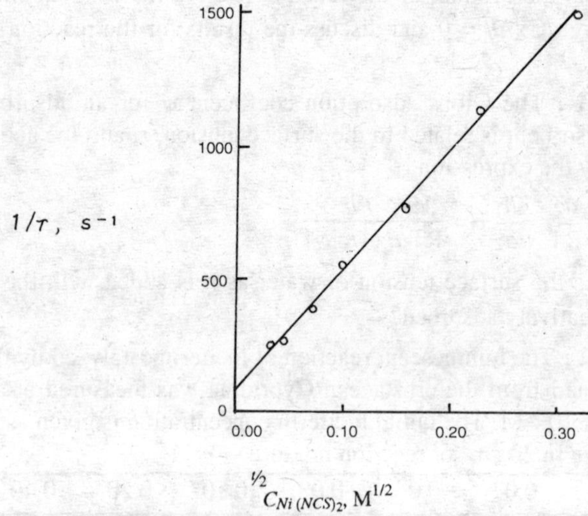

in which the equilibrium is disturbed. It is found that the relaxation time for the return to equilibrium at 25°C is 37μs. Calculate the rate coefficients for the forward and backward reactions.

Promblem 25 : The following mechanism is proposed for the photodimerization of substance A:

$A + h\upsilon \xrightarrow{k_1} A^*$

$A^* + A \xrightarrow{k_2} A_2$

$A^* \xrightarrow{k_3} A + h\upsilon^1$

Here, h = Planck's constant and

υ = Frequency of incident light

Derive the expression for the quantum yield of A_2.

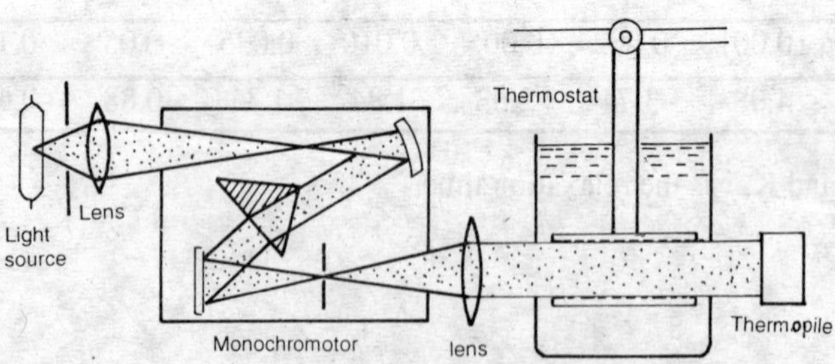

Arrangement for Photochemical investigations

Problem 26 : The rate of a first order reaction increases from $r = 1.5 \times 10^{-2}$ sec^{1} to $r_c = 4.6\ sec^{-1}$ at 260°C when a catalyst is added to the reaction. Calculate the decrease in the activation enthalpy H, assuming S°, the activation entropy, is not affected by the catalyst.

Problem 27 :

	Without Catalyst (k), sec^{-1}	With Catalyst (k_c) sec^{-1}
$T_1 = 473°$ K	1.76×10^{-2}	6.10
$T_2 = 573°$K	0.804	88.1

The reaction rate of a first-order reaction at a constant pressure of 1 atm is shown in the table above. Calculate the value of H and S° with and without catalyst, assuming that they do not change with temperature.

Problem 28: The decomposition of benzoyl peroxide in diethyl ether, a first-order reaction, is 75.2 percent complete in 10 min at 60° C.

Calculate the rate constant for the reaction.

Problem 29: In a study of the reaction between

Br^- and ClO^- $ClO^- + Br^- \rightarrow BrO^- + Cl^-$

100 ml of 0.1 N NaClO, 48 ml of 0.5 N NaOH, and 21 ml of distilled water were mixed and immersed this mixture in a themostat at 25°C. To this mixture 81 ml of a 1 percent KBr solution, also at 25°C, was added. At certain intervals, t, samples were withdrawn for the determination of BrO^-, The resutls of the analyses are given below:

t, min	x,conc of BrO^- in moles $liter^{-1} \times 10^2$
0.0	0.0
3.65	0.0560
7.65	0.0953
15.05	0.1420
26.00	0.1800
47.60	0.2117
90.60	0.2367

The concentration of NaClO in the reaction mixture at $t = 0$ was 0.003230 *M*, and that of KB*r* was 0.002508 *M*. The *p*H was 11.28. Determine the reaction order and the value of the reaction rate constant.

Problem 30: To investigate the decomposition of oxalic acid in concentrated H_2SO_4 at 50°C, Lichty [*J. Phys, Chem.*, 11, 225 (1907)] prepared a 1/40 *M* solution of oxalix acid in 99.5 percent H_2SO_4, then removed aliquots at various reaction times *t*, and then determined the volumes υ of a potassium permanganate solution required to react with 10 ml portion. The results are given below:

t, min	υ, ml
0	11.45
120	9.63
240	8.11
420	6.22
600	4.79
900	2.97
1440	1.44

Determine the reaction order with respect to oxalic acid and evaluate the specific rate constant.

Problem 31: The kinetics of the reaction between sodium thiosulfate and *n*-propyl bromide

$$S_2O_3^{2-} + RBr \longrightarrow RSSO_3^- + Br^-$$

was investigated at 37.50°C by Crowell and Hammett [*J. Am. Chem. Soc.*, 70, 3444 (1948)]. The concentration of $S_2O_3^{2-}$, initially 0.100 *M*, was determined at various intervals *t* by titration with υ ml of 0.02572 *N* iodine solution per 10.02 ml sample of reaction mixture. Thiosulfate was present in excess. The results are given below:

t, sec	υ, ml
0	37.63
1110	35.20

2010	33.63
3192	31.90
5052	29.86
7380	28.04
11232	26.01
78840	22.24

Evaluate the second-order rate constant κ.

Problem 32: The rate of the reaction

$$(C_2H_5)_3N + CH_3I \longrightarrow [(C_2H_5)_3 \overset{+}{N}(CH3)I^-$$

in nitrobenzene at 25°C was investigated by Brown and Eldred [*J. Am. Chem. Soc.*, 71, 445 (1949)] who found that:

t ,sec	υ, moles liter^{-1}
1200	0.00876
1800	0.01066
2400	0.01208
3600	0.01392
4500	0.01476
5400	0.01538

where t is the reaction time in seconds and x is the moles liter^{-1} of $(C_2H_5)_3N$ or CH_3I reacted at time t. The initial concentrations of amine and alkyl iodide were each 0.0198 moles liter^{-1}. The reaction is second order. Find the rate constant.

Problem 33: Calculate the order of the rate-controlling step in the pyrolysis of diborane (B_2H_6, g) at 100°C from the following data on the rate of pressure increase in the system as a function of B_2H_6 concentration [Bragg, McCarty, and Norton, *J. Am. Chem. Soc.*, 73, 2134 (1951)]:

Concentration of B_2H_6, moles liter^{-1} × 10^2	Rate of pressure increase, moles liter^{-1} × 10^4
2.153	7.4
0.433	0.73

Problem 34: The kinetics of the catalytic decomposition of NH_3 into the

elements on a hot tungsten filament at 1100ÉC was investigated by Kunsman [*J. Am. Chem. Soc.*, 50, 2100 (1928)] who found that the times $T_{1/2}$ required for half of the NH_3 to decompose with no N_2 or H_2 present at the start depended on the initial NH_3 pressures as follows:

p, torr	t1/2 , min
265	7.6
130	3.7
58	1.7

Determine (a) the order of the reaction, and (b) the rate constant.

Problem 35: The kinetics of the reaction of methanol with 3, 3′-dicarbazylphenylmethyl ion (a green dye) in alkaline water-acetone solution was investigated spectrophotometrically at 25°C by Branch and Tolbert [*J. Am. Chem. Soc.*, 69, 523 (1947)]. They found that for a solution initially 0.25 *M* in methanol, the difference $A - A_\infty$ between the absorbance of the solution at time *t* (A) and the absorbance of the solution when the reaction reached equilibrium (A_∞) at 730 nm varied with time *t* as follows, when the NaOH concentration is maintained at 2.03×10^{-2} *M* (selected data):

t, min	$A - A_\infty$
2.7	0.562
8.7	0.243
14.7	0.111
21.7	0.045
33.7	0.010

Show that the reaction is first order with respect to the 3,3′-dicarbazylphenylmethyl ion and evaluate the rate constant. For light passed through a sample, let *P* be transmitted power and P_0 be incident power. Then absorbance, *A*, is given by

$$A = \log \frac{P_0}{P}$$

Absorbance is directly proportional to the concentration of the absorbing species.

Problem 36 : The rate of inversion of sucrose in the presence of HCl was studied by Fales and Morrell [*J. Am. Chem. Soc.*, **44**, 2071 (1922)] who obtained the following data :

t, time elapsed, sec	α, observed angle of rotation, deg
0	11.20
1035.00	10.35
3113	8.87
4857	7.64
9231	5.19
12834	3.61
18520	1.60
26320	-0.16
32640	-1.10
76969	-3.26
∞	-337

Determine the order of the reaction and evaluate the specific rate constant.

Problem 37: In order to study the kinetics of the reaction

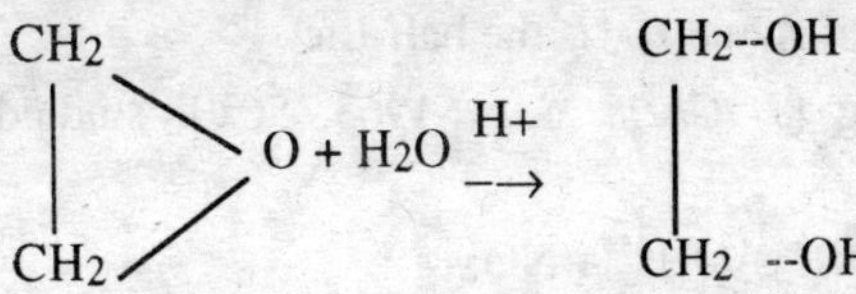

a solution initially 0.12 *M* in ethylene oxide and 0.007574 *M* in perchloric acid was prepared and the progress of the reaction was followed dilatometrically (that is, by measuring the volume of the solution as a function of time). The following data were obtained at 20°C:

t, min	Dilatometer reading
0	18.48
30	18.05
60	17.62
135	16.71
300	15.22
∞	12.29

Determine the order of the reaction with respect to ethylene oxide and evaluate the specific rate constant.

Problem 38 : Show that the third-order dependence of the reaction

$2NO + H_2 \rightarrow 2NOH$

can be accounted for by either of the mechanisms

(a) $NO + H_2 \overset{k}{\Leftrightarrow} NOH_2$

$NOH_2 + NO \xrightarrow{k} 2\,NOH$

(b) $2NO \overset{k'}{=} N_2O_2$

$N_2O_2 + H_2 \xrightarrow{k'} 2\,NOH$

no step of which is termolecular.

Problem 39 : Show that for a

(a) first-order reaction, $A \rightarrow$ products, $t_{1/2} = 0.693/k$

(b) second-order reaction, $2A \rightarrow$ products, $t_{1/2} = 1/ka$

(c) third-second reaction, $3A \rightarrow$ products, $t_{1/2} = 3/2/ka^2$

where a is the initial concentration of A and $t_{1/2}$ is the half-life.

Problem 40: Lalor and Lang [*J. Chem. Soc.,* 1963, 5620] studied the kinetics of the reaction

$Co(NH_3)_5NO_2^{2+} + OH^- \rightarrow Co(NH_3)_5OH^{2+} + NO_2^-$

and found the following dependence of the rate constant k on ionic strength I:

I	$5 + \log k$
2.34	1.7640
5.61	1.7130
8.10	1.6800
11.22	1.6467
11.73	1.6418
16.90	1.5990

Evaluate k_0, the ratre constant at zero ionic strength.

Problem 41: It is often said that the rate of a typical chemical reaction doubles with every 10-degree rise in temperature. What must be the activation energy of a reaction for which this statement is true in the vicinity of 300°K?

Problem 42 : The rate of the gas reaction

$N + O_2 \rightarrow NO + O$

has been studied [Wilson, *J. Chem. Phys.*, **46,** 2017 (1967) by generating N atoms in a flow system by a microwave discharge and monitoring their concentration by electron-spin resonance. The following rate constants were found:

T, °K	*k*, $cm^3\ mole^{-1}\ sec^{-1}$
586	$(1.63 \pm 0.05) \times 10^{10}$
910	$(1.77 \pm 0.08) \times 10^{11}$

Calculate the activation energy of this reaction. Obtain two values corresponding to the upper and lower limits permitted by the given uncertainties in the data.

Problem 43: In an investigation of the kinetics of the reaction

$MnO_4^- + Cr^3 \rightarrow CrO_4^{2-} + Mn^{4+}$ at 25.1 °C

Fales and Roller [*J. Am. Chem. Soc.*, **51,** 345 (1929)] measured the times required to carry the reaction to various degrees of completion, first as a function of CrO_4^{2-} concentration and then as a function of MnO_4^- concentration. In a set of three experiments, the results were as follows:

Experiments No.	I	II	III
Relative MnO_4^- conc.	1	2	1
Relative CrO_4^{2-} conc.	1	1	0.5
$Cr_2O_7^{2-}$, ml (0.1 *N*)			
0.1	23 min	11 min	40 min
0.2	36	18	70
0.4	58	32	113
0.6	77	44	148

In each case after definite intervals of time 25 ml aliquots of the reaction mixtures were removed and added rapidly to a solution consisting of 60 ml of 1.2*M* H_2SO_4 and 25 ml of 0.5 *M* K br. Free bromine was instantly liberated by the reaction

$2MnO_4^- + 10Br^- + 16H^+ \rightarrow 2Mn^{2+} + 5Br_2 + 8H_2O$

and was extracted by CS_2. The $Cr_2O_7^{2-}$ that had been formed by the reaction

$2CrO_7^{2-} + 2H^+ \rightarrow Cr_2O_7^{2-} + H_2O$

was titrated with 0.01 *N* $FeSO_4$ in accordance with the reaction

$Cr_2O_7^{2-} + 6Fe^{2+} + 16H^+ + 2SO_4^{2-} \rightarrow 2Cr^{3+} + 6Fe^{3+} + 2HSO_4^- + 7H_2O$

The number of milliliters of 0.01 *N* $Cr_2O_7^{2-}$ present in the mixture at the indicated reaction time is given in the table. This volume may be thought of simply as a number that is proportional to the concentration of CrO_4^{2-} in the reaction mixture at the indicated time. What is the reaction order (a) with respect to Cr^{3+}, (b) with respect to MnO_4^-?

Problem 44: The kinetics of the hydrolysis of *t*-butyl chloride (net reaction: $(CH_3)_3CCl + H_2O \rightarrow (CH_3)_3COH + H^+ Cl^-$)

at 25°C was investigated by Swain and Ross [*J. Am. Chem. Soc.*, 68, 658 (1946] by means of a concentration cell of the type Ag | AgCl | A || B | AgCI } Ag. At zero time both half cells contained a 95 percent water--5 percent acetone solution which was 0.144 *M* in sodium perchlorate; but, in addition, A contained 0.001 *M* *t*-butyl chloride. The rate measurements were then carried out by running small measured additions of 0.0499 *M* HCI into cell B every few seconds and the time was recorded at each moment when a galvanometer placed between the two electrodes registered zero for each addition. The follwoing data were obtained for one such sereis of runs:

t, sec	ml HCl in B
30	0.90
44	1.10
62	1.30
83	1.40
111	1.50
∞	1.61

Show that the hydrolysis reaction is first order, and evaluate the rate constant.

Problem 45 : In an investigation of the kinetics of the alkaline hydrolysis of of ethyl acetate, Potts and Amis [*J. Am. Chem. Soc.*, **71**, 2112 (1949)] carried out the following experiment. 300 ml of water at 0°C was pipetted into a 600-ml flask and 100.00 ml of 0.0500 *M* ethyl acetate was added at the same temperature. The flask was stoppered and allowed to reach thermal equilibrium in a thermostat. The hydrolysis reaction was then initiated by additing 100.00 ml of 0.01000 *M* NaOH, also at 0°C, and shaking vigorously. If we assume that the volumes of the starting solutions are strictly additive on mixing, the resulting solution was initially 0.0100 *M* in ethyl acetate and 0.0200 *M* in NaOH. At various intervals, *t*, 50.00-ml samples of the reaction mixture were withdrawn, drained into a measured excess of 0.0200 *M* HCl at 0°C (v_{HCl}), and titrated at 0°C with 0.0200 *M* NaOH in the reaction mixture at equilibrium time ($t = \infty$) was determined on samples permitted to stand at the reaction temperature until no further change in the composition of the mixture as a function of time could be detected. The results were as follows:

t, min	v_{HCl}, ml	v_{NaOH}, ml
0	60.0	11.50
15	50.0	7.35
30	50.0	11.20
45	50.0	13.95
60	50.0	15.95
75	50.0	17.55
∞	50.0	24.00

Determine the order of this reaction and the rate constant *k*.

Problem 46 : The reaction rate of the photochemcial decomposition of $K_2S_2O_8$

$$K_2S_2O_8 + H_2O \xrightarrow{h\nu} 2KHSO_4 + \frac{1}{2}O_2$$

in the presence of 0.1 *M* K_2SO_2 and 0.28 *M* KOH was investigated by Morgan and Crist [*J. Am. Chem. Soc.*, **49**, 16 (1927)] who titrated aliquots of the reaction mixture removed at the times indicated and obtained the following data:

time, min	ml $FeSO_4$, consumed
0	9.25
150	8.75
360	8.24

540	7.86
720	7.51
940	7.07
1160	6.64
1340	6.27

0.0500 *M* $FeSO_4$ was the reagent and the titration reaction was

$K_2S_2O_8 + FeSO_4 \rightarrow Fe_2(SO_4)_3 + K_2SO$

The initial concentration of $K_2S_2O_8$ was 0.100 *M*. Find the order of the reaction with respect to $K_2S_2O_8$.

Problem 47: In the presence of sodium methoxide in methanol solution, chloroform reacts first in accordance with the equation

$CHCl_3 + OMe^- \rightleftharpoons CCl_3^- + MeOH$ (*fast, K* » 1)

and then

$CCl_3^- \rightarrow CCl_2 + Cl^-$ (*slow*)

followed by $CCl_2 \rightarrow$products (fast). The slow middle step is believed to be the rate-controlling one. In an experiment to determine the order of the slow step with repect to CCl_3^-, the amount of Cl^- produced in time *t* at 59.7 °C was determined by titration with 0.0100 *N* $AgNO)_3$, ν ml being rquired [O'Ferral and Ridd, *J. Chem. Soc.*, **1963**, 5035]. The results were as follows:

t, min	ν, ml
0	1.71
4	3.03
9	4.49
15	5.97
22	7.39
30	8.87
41	10.48
50	11.7
∞	15.98

The initial $[CHCl_3]$ was 9.95×10^{-3} *M* and the mean [NaOMe] during the reaction was 1.10 ± 0.01 *M* (a large excess, practically constant during the run). Show that the reaction is first order with respect to CCl_3^- and evaluate the rate constant.

Problem 48: The kinetics of the oxidation of oxalic acid by ceric sulfate was investigated by Ross and Swain [*J. Am. Chem. Soc.*, **69,** 1325 (1947)] by means of a concentration cell of the type Pt | S_1 || S_2 | Pt. Initially S_1 and S_2 contained equal amounts of sulfuric acid, ammonium bisulfate (added to maintain ther mixture at a constant ionic strength of 2.0 *M*), and ceric sulfate. At zero time a standard solution of oxalic acid was added to S_1 and equal volume of water to S_2. The rate measurements were then carried out by adding measured excess of ferrous sulfate to solution S_2 and noting the times at which a galvanometer placed between the two Pt electrodes registered zero for each addition. The following data were obtained for one series of runs using 0.00253 *M* $Ce(SO_4)_2$, 0.00125 *M* $H_2C_2O_4$, AND 2.*M* H_2HO_4:

t, min	ml $FeSO_4$ (*cumulative*)
0.31	7.00
0.51	9.00
0.65	10.00
0.86	11.00
1.01	11.50
1.15	12.00
1.36	12.50
1.58	13.00
2.19	13.80
∞	16.50

Show that the reaction between oxalic acid and ceric sulfate is second order, and evaluate the rate constant.

Problem 49: The thermal decomposition of arsine on glass,
$2AsH_3(g) \rightarrow 2As(s) + 3H_2(g)$
was investigated by K. Tararu [*J. Phys. Chem.*, **59,** 777 (1955)] who found that the total pressure of the system varied with time as indicated below at 350°C:

Time, hr	Total pressure, cm
0	39.2
4.33	40.3
16	43.65

25.5	45.35
37.66	48.05
44.45	48.85

(a) Determine the order of the reaction with respect to arsine and (b) evaluate the specific rate constant.

Problem 50: In an investigation of the hydrolysis of methyl tosylate (methyl p-toluenesulfonate), Swain and Morgan [J. Org. Chem.,29, 2097 (1967)] obtained the following spectrophotometric data at 261 nm and 25°C on a solution in water initially aa1.22 X 10^{-3} M with respect to methyl tosylate.

Time, hr	Absorbance
0.0	0.791
5.0	0.746
9.0	0.709
18.0	0.645
24.0	0.611
32.0	0.570
44.0	0.526
55.0	0.498
67.0	0.471
290.0	0.420

Assuming that methyl tosylate is the only absorbing species at this wavelength, show that the reaction is first order with respect to methyl tosylate and evaluate the rate constrant in sec^{-1}.

Problem 51: A gas that undregoes a unimolecular reaction is mixed with an inert gas. Using the Lindemann theory, find the rate of the reaction in terms of the concentrations of the reacting gas and the ilnert gas and of the rate constants of the elementary reactions involved.

Problem 52: The following mechanism for lthe thermal decomposition of ozone has been proposed:

$$O_3 \xrightarrow{k_1} O_2 + O$$

$$O + O_2 \xrightarrow{k_{-1}} O_3$$

(a) By the method of the stationary state obtain an expression for the rate of decomposition in terms of the rate constants of these three reactions and the concentrations of O_2 and O_3.

(b) The reaction is found experimentally to be of the second order with respect to O_3, and of the -1 order with respect to O_2. State the relations of magnitude that must exist among the concentrations and rate constants in order that your expression may agree with these facts.

Problem 53: The following mechanism (i) for the reaction

$H_2(g) + I_2(g) \rightarrow 2HI(g)$

was generally accepted until 1967:

$H_2 + I_2 \rightarrow 2HI$ Rate constant: k

In that year it was shown (Sullivan, *J. Chem. Phys.*, **46**,73) that the reaction probably proceeds by the mechanisum (ii):

		Rate constant
$l_2 \rightarrow 21$	(1)	k_1
$21 \rightarrow l_2$	(2)	k_2
$21 + H_2 \rightarrow 2H1$	(3)	k_3

(a) Obtain the rate law for d[HI]/dt predicted by each of the mechanisms (i) and (ii). Use the steady state method where appropriate.

(b) Obtain the rate law predicted by mechanism (ii) under the assumption that reactions (1) and (2) comprise an equilibrium, with K being the equilibrium constant for the reaction $l_2 \rightleftharpoons 21$. Under what conditions do the results of (a) and (b) for mechanism (ii) agree?

(c) By producing I atoms photochemically in known concentration it is possible to measure k_3,

T(°K)	417.9	520.1
k_3(liter^2 mole^{-2}sec^{-1})	1.12×10^5	4.0×10^5

(k_3 is defined with reference to $-d[H_2]/dt$.) Calculate the activation energy for reaction (3).

Problem 54: The following mechanism has been proposed by Anderson and Freeman [*J.Phys. Chem.*,**65**,1648 (1961)] for the reaction between sodium nitrite and oxygen:

$$NO_2^- + O_2 \xrightarrow{k_1} NO_3^- + O$$

$$O + NO_2 \xrightarrow{k_2} NO_3^-$$

$$O + O \xrightarrow{k_3} O_2$$

(a) Show that

$$\frac{d[NO_3-]}{dt} = k_1[NO_2-][O_2]\left(1 + \frac{k_2[NO_2-]}{k_3[O] + k_2[NO_2-]}\right)$$

(b) Show that the expression given in (a) reduces to

$$\frac{d[NO_3-]}{dt} = 2k_1[NO_2-][O_2]$$

if we assume that reaction (3) is much slower than reaction (2).

Problem 55: The decomposition of nitramide in aqueous solution,

$O_2NNH_2 \rightarrow N_2O + H_2O$

is first order with respect to nitramide. It is believed that the reaction proceeds in three steps,

$O_2NNH_2 + B \rightarrow BH^+ + O_2NNH-$ (1)

$O_2NNH- \rightarrow N_2O + OH-$ (2)

$OH- + BH^+ \rightarrow B + H_2O$ (3)

B is some base, for example, acetate ion.

Two hypotheses have been proposed.

(a) Reaction (1) is fast in both directions with equilibrium constant K; (2) is slow and rate determining with rate constant k_2.

(b) Reaction (1) is rate determining with rate constant k_1 and (2) is much faster.

Ascertain whether either, or both, of these hypotheses predicts the first-order rate law correctly. (Assume that activities are equal to concentrations.) If both are consistent with the facts, suggest a further experiment that might serve to exclude one or the other.

Problem 56: The following mechanism has been proposed for the radiation-induced decomposition of several inorganic nitrates at room temperature [Chen and Johnsom, *J. Phys. Chem.*, 66, 2249 (1962)].

$NO_3- \xrightarrow{k_1\phi} NO_2- + O$ (1)

$O + NO_2- \xrightarrow{k_2} NO_3-$ (2)

$O + NO_3- \xrightarrow{k_3} NO_2- + O_2$ (3)

The symbol Ørepresents the incident dose of the radiation and indicates that ionizing radiation instead of light) was used. (a) Making the steady state assumption about O, derive the rate expression for the formation of NO_2 -.

(b) Integrate the rate expression obtained in (a) assuming that the concentration of NO_3^- remains constant.

Problem 57: The following mechanism for the homogeneous pyrolysis of methane has been proposed:

$$CH_4 \xrightarrow{k_1} CH_3 + H$$

$$CH_3 + CH_4 \xrightarrow{k_2} C_2H_6 + H$$

$$H + CH_4 \xrightarrow{k_3} CH_3 + H_2$$

$$H + CH_3 + M \xrightarrow{k_4} CH_4 + M$$

M is any molecule (perhaps CH_4 or C_2H_6) that can carry away the energy of recombination of H and CH_3.

Deduce the rate law for the formation of C_2H_6 predicted by this mechanism. The concentrations of CH_3 and H, which are present in very small, steady concentration, should not apear in the result. [Cagle, *J. Chem. Phys.*,25,1300 (1956)].

Problem 58: The hypothetical reaction A→ B is of the -1 order; that is,$-dc/dt = kc^{-1}$, where c= [A].

(a) Obtain an equation for c as a function of t, k, and the initial concentration c_0.

(b) Find the time required for the concentration to fall to 10 percent oif lits initial value, in terms of k and c_0.

(c) Does this reaction ever reach completion?

Does a first-order reaction ever reach completion? Explain any difference between the two cases.

Problem 59: The following mechanism has been proposed for the reaction $NO_2 \rightarrow NO_2 + Cl_2$

$$NO_2Cl \xrightarrow{k_1} NO_2 + Cl$$

$$NO_2Cl + Cl \xrightarrow{k_2} NO_2 + Cl_2$$

(a) Find the rate law predicted by this mechanism. Assume that Cl is present in a small, steady concentration.

(b) At NO_2Cl concentration below 10^{-4} mole liter^{-1} the reaction becomes second order in NO_2Cl. Write a modified version of the above mechanism that is consistent with this rate low and explain why the modification is necessary only at low concentration.

Problem 60: H. Johnson, L. Foering, and R. J. Thompson [J. *Phys. Chem.*,

57, 390 (1953)] propose that nitric acid decomposes by the following mechanism:

$$HNO_3 \xrightarrow{k_a} HO + NO_2$$

$$HO + NO_2 \xrightarrow{k_b} HNO_3$$

$$HO + HNO_3 \xrightarrow{k_c} H_2O + NO_3$$

(a) Making the steady state assumption about HO, show that the rate expression for the decomposition of HNO_3 is given by

$$\frac{d[HNO_3]}{dt} = -2k_a [HNO_3] \cdot \frac{1}{1 + (k_a [NO_2])/(k_c[HNO_3])}$$

(b) If NO_2 is consumed by the fast reaction

$$NO_3 + NO_2 \xrightarrow{k_1} NO_2 + O_2 + NO$$

what form does the above rate expression reduce to?

Problem 61 : Making the steady state assumption about HO_2, show that the following mechanism for lthe reduction of neptunium (VI) by hydrogen peroxide leads to the rate expression

$$\frac{d[N_pO_2^+]}{dt} = \frac{2Kk_1 [N_pO_2^{2+}] [H_2O_2]}{[H+] \{1 + k_3[N_pO_2^{2+}])/(k_2 [N_pO_2^{2+}]\}}$$

[Zielen, Sullivan, Cohen, and Hindan, *J. Am. Chem. Soc.*, **80,** 5632 (1958).]

$$H_2O_2 \overset{k}{=} H^+ + HO_2 -$$

$$HO_2 - + Np\,O_2^{2+} \xrightarrow{k_1} NpO_2+ + HO_2$$

$$HO_2 + Np\,O_2^{2+} \xrightarrow{k_2} NpO_2 + O_2 + H^+$$

$$HO_2 + NpO_2+ \xrightarrow{k_3} NpO_2^{2+} + HO_2 -$$

Problem 62 : Assume that the thermal decomposition of gaseous N_2O_5 proceeds by the following mechanism:

$$N_2O_5 \xrightarrow{k_1} NO_2 + NO_3$$

$$NO_2 + NO_3 \xrightarrow{k_{-1}} N_2O_5$$

$$NO_2 + NO_3 \xrightarrow{k_2} NO_2 + O_2 + NO$$

$$NO + N_2O_5 \xrightarrow{k_3} 3NO_2$$

(a) By the method of the steady state applied to NO_3 and NO find the rate law that this mechanism predicts.

(b) It is believed that the activation energy of reaction (-1) is close to 0 but that the activation energy of reaction (2) is about 5 kcal. Maing reasonable simplifying assumptions, estimate roughly the ratio k_2/k_{-1} at 320°K. Explain why your assumption are reasonable.

(c) Show how essentially the same rate law obtained in (a) can be deduced by treating reaction (1) and (-1) as an equilibrium. Under what condition is this approach justified?

Problem 63: The following mechanism for the formation of phosgene, $COCl_2$, has been proposed :

$$Cl_2 \xrightarrow{k_1^*} 2Cl \qquad \text{* for } d[Cl_2]/dt$$

$$2Cl \xrightarrow{k_2^*} Cl_2$$

$$Cl + CO \xrightarrow{k_3} COCl$$

$$COCl \xrightarrow{k_4} Cl + CO$$

$$COCl + Cl_2 \xrightarrow{k_5} COCl_2 + Cl$$

(a) Use the steady state approximation to find the rate equation implied by this mechanism for the formation of $COCl_2$. Assume that CI and COCI are present in small, steady concentrations. The rate equation may contain any or all of the constant k_1 ,...., k_2 and the concentrations of all species other than Cl or COCl.

(b) Assume that reactions (1)-(4) are much faster than reaction (5). Show how the rate equation can be simplified in this case.

(c) Show that the result of (b) can also be obtained by treating reaction (1)-(2) and reactions (3)-(4) as equilibria. Explain why the assumptions of (b) and (c) lead to the same result.

Promblem 64: If two first-order reactions are operating in succession, viz.

$$A \xrightarrow{k_1} B \xrightarrow{k_2} C$$

(starting with A but no B or C), prove that the concentration of the intermediate B reaches a maximum and then declines, evaluate the time at which the maximum concentration of B occurs, and evaluate the maximum concentration of B. (Note: This problem requires a little knowledge of linear differential equations beyond the ususal introductory course in calculus. Remember that $e^{a \ln b} = b^a$. The case $k_1 = k_2$ requires separate consideration.)

PHOTOCHEMISTRY

Problem 1: On the basis of the following data, selected from the work of Discher, Smith, Lippman, and Turse [*J, Phys. Chem.*, **67,** 2501 (1963)], evaluate the quantum efficiency of the uranyl oxalate system at each wavelength.

Wavelength, nm	Fraction of oxalate decomposed	Molecules decomposed $\times 10^{-18}$	Photons absorbed $\times 10^{-18}$
365.5	0.0592	5.18	10.58
365.5	0.0498	4.32	8.93
435.8	0.0242	2.10	3.64
435.8	0.0208	1.79	3.10

Problem 2 : In an experiment to determine the concentrations of the two substances A and B in a certain solution simultaneously by spectrophotometry the following data were obtained:

	[A] mole liter^{-1}	[B] mole liter-1	Percent transmitted at $\lambda = 400$ mm	percent transmitted at $\lambda = 500$ mm
Solution 1	0.001	0	10	60
Solution 2	0	0.005	80	20
Solution 3	unknown	unknown	40	50

Determine the concentrations of A and B in solution 3.

Problem 3 : In order to determine the dissociation constant *K* of the indicator *m*-nitrophenol (HIn), Robinson and Peiperl [*J. Phys. Chem.*, **67,** 2860 (1963)] prepared a 6.36×10^{-4} *M* solution of it and carried out the following spectrophotometric measurements on the solution at 25°C and 390 nm. In one experiment the solution was made strongly acidic and the absorbance A_1 was found to be 0.142. Under these conditions essentially all of the indicator is in the HIn form. In another experiment the solution was made strongly basic and the essentially all of the indicator is in the In- form. Then in a series of additional experiments the *p*H of the indicator solution was adjusted with a buffer mixture of ionic strength *I* and the absorbances *A* were measured. The results were as follows:

I	*p*H	*A*
0.10	8.321	0.527
0.08	8.302	0.518
0.06	8.280	0.505
0.04	8.251	0.493
0.02	8.207	0.470

Evaluate the apparaent *p*K (in terms of concentrations) of the indicator at each of the ionic strengths listed.

Problem 4 : The percent transmittance (percent *T*) of a series of solutions of the acid-base indicator bromophenol blue were determined at 590 nm, and the results are as follows (Labowitz, unpublished data):

Relative concentration	Buffered at *p*H 4.39 percent T	Strongly basic percent T
0.8	20.0	5.4
0.6	29.0	11.0
0.4	44.2	22.7
0.2	66.8	46.4

At the indicated wavelength, the base form absorbs strongly, but the absorbance of the acid form is negligible. Calculate the pK_a of this indicator.

Problem 5 : The following mechanism [Buxton and Walker, *J. Phys. Chem.*, **67,** 2835 (1963)] is proposed for the photochemical decomposition of hydrogen peroxide in the presence of carbon monoxide. Making the steady state assumption about OH and CO_2H, and HO_2 , obtain the rate law for $d[H_2O_2]/dt$. Let ϕ represent the primary quantum efficiency for H_2O_2 photolysis and Ia represents einsteins absorbed per liter per second.

$$H_2O_2 + h\nu \longrightarrow 2OH$$

$$OH + CO \xrightarrow{k_1} CO_2H$$

$$CO_2H + H_2O_2 \xrightarrow{k_2} CO_2 + H_2O + OH$$

$$H_2O_2 + OH \xrightarrow{k_3} H_2O + HO_2$$

$$2HO_2 \xrightarrow{k_4} O_2 + H_2O$$

Problem 6: The following alternative mechanism is proposed for the photolytic decomposition of hydrogen peroxide in the presence of carbon monoxide [Buxton and Walker, *J. Phys. Chem.*, **67,** 2835 (1963)]. Making the steady state assumption about OH and CO_2H, evaluate $d\,[H_2O_2]/dt$ as in the preceding problem.

$$H\,O_2 + h\nu \rightarrow 2OH$$

$$OH + CO \xrightarrow{k_1} CO_2H$$

$$CO_2H + H_2O_2 \xrightarrow{k_2} CO_2 + H_2O + OH$$

$$2CO_2H \xrightarrow{k_5} \text{products}$$

Problem 7: In a spectrophotometric analysis, at what value of T (fractional transmittance) will the relative error in concentration ($\Delta c/c$) be minimal for a given error in the measurement of T?

Problem 8: (a) In the photochemical gas reaction $2HI \rightarrow H_2 + I_2$ the quantum yield is approximately 2. Find the number of grams of HI decomposed per calorie of radiant energy absorbed when HI is irradiated with monochromatic radiation with wave-length 2070 Å.

(b) Is the following mechanism in agreement with the observed quantum yield ? Explain.

(1) $HI + h\nu \rightarrow H + I$

(2) $H + HI \rightarrow H_2 + I$

(3) $2I \rightarrow I_2$ (on wall or third body)

(c) In the reaction $H_2 + Cl_2 \rightarrow 2HCl$, which has a mechanism similar in some ways, quantum yields of 10^4 to 10^6 are obtained. Explain the great difference between the quantum yields in the two reactions, with reference to relevant energies of activation, bond energies, or other data. The energy of the H-I bond is 71 kcal $mole_{-1}$. That of I-I is 36 kcal $mole^{-1}$. Energies of activation are

$Cl + H_2 \rightarrow HCl + H$	8 kcal
$H + Cl_2 \rightarrow HCl + Cl$	2 kcal
$H + HI \rightarrow H_2 + I$	1.5 kcal

Problem 9 : The photolysis of O_3 in liquid A*r* solution at 87°K has been studied [De Mare and Raper, *J. Chem. Phys.*, **44**, 1780 (1966)]. Radiation of wavelengths 2537 Å and 3130 Å was used. The primary quantum yield, ϕ, is so defined that the rate of decomposition of O_3 by the reaction

$$O_3 + h\nu \rightarrow O_2 + O^* \qquad (1)$$

is νI_a mole liter^{-1} sec^{-1}, where I_a is einsteins absorbed per liter per second and O* is an O atom in an excited electronic state (^{1}D). The reactions following reaction (1) were assumed to be

$$O^* + O_3 \rightarrow 2O_2 \qquad (2)$$

$$O^* \rightarrow O \qquad (3)$$

$$O + O_2 + M \rightarrow O_3 + M \qquad (4)$$

where M is any molecule that can carry away the energy liberated in reaction (4). The overall quantum yield ϕ is so defined that the net rate of O_3 disappearance is $-d[O_3]/dt = \phi I_a$.

(a) On the basis of the mechanism given obtain an expression for ϕ in terms of ϕ, rate constants, and concentrations of stable species (O_2, O_3).

(b) The experimental results for ϕ at 2537° fit an equation of the form ϕ-1 =0.538 + 0.81 $[O_3]^{-1}$. Find ϕ and any available information about the rate constants.

Problem 10: The kinetics and mechanism of the vapour phase photochlorination of trifluoroethylene

$$CF_2 = CFH + Cl_2 \xrightarrow{h\nu} CF_2Cl = CFHCl \text{ (or } TfCl_2)$$

were investigated at room temperature by Bunbury, Lacher, and Park [*J. Am. Chem. Soc.*, **80,** 5104 (1958)], who observed that

I. There is practically no dark reaction.

II. In the range 25 to 400 torr the reaction rate is independent of the olefin pressure.

III. When the chlorine pressure is 90 torr, the reaction rate ($-dP_{Cl_2}/dt$) varies with the light intensity I_0 in the manner indicated below.

I_0 arbitrary units	$(-dp_{Cl_2}/dt) \times 10^3$ torss sec^{-1}
0.0144	2.2
0.122	6.1
0.421	9.5
1.88	17.2
1.90	18.0
1.90	18.4
1.93	19.6

IV. When the light intensity is constant, the reaction rate changes with Cl_2 pressure in the manner indicated below.

P_{Cl_2}, torr	$(-dP_{cl2}/dt) \times 10^3$ torr sec^{-1}
9.1	1.5
22.2	6.98
35	11.66
45	17.3
94	47.5
137.2	77.5
180.7	101

V. The reaction is completely stopped when traces of NO are added to the system.

(a) Referring to the data in III above it is seen that when the Cl_2 pressure is constant, the reaction rate is proportional to I_0^n. Evaluate *n*.

(b) Referring to the data in IV above it is seen that when the light intensity is constant, the reaction rate is proportional to $P_{cl_2}^m$. Evaluate *m*.

(c) On the basis of your findings in (a) and (b) write the empirical rate equation for the overall reaction.

(d) The following mechanism (where Tf stands for $CF_2 = CFH$) was proposed by the investigators. Apply the steady state approximation method and show that the proposed mechansim is consistent with the empirical rate law. Simplify the results as much as possible.

$$Cl_2 + h\nu \text{ (incident light)} \rightarrow 2Cl \quad (1)$$

$$Cl + Tf \rightarrow TfCl \quad (2)$$

$$TfCl + Cl_2 \rightarrow TfCl + cl \quad (3)$$

$$TfCl + TfCl \rightarrow \text{products} \quad (4)$$

(e) In equation (4) the products were not identified. Under what conditions is it legitimate to disregard the identity of products?

(f) If you wished to show experimentally that reaction (3) is not reversible, how would you proceed?

(g) What is the significance of observation V?

Part IV
SOLUTIONS TO PROBLEMS

CHEMICAL KINETICS

Solution 1 : For equation (3) of problem, the rate equation is

$$\frac{d\,[C_2H_6]}{dt} = k_3\,[C_2H_5]\,[H_2]$$

Assume that H and C_2H_2 reach steady state concentrations.

$$\frac{d\,[H]}{dt} = 2k_1\,[Hg]\,[H_2] - k_2\,[H]\,[C_2H_4] + k_3\,[C_2H_2]\,[H_2]$$
$$-2k_4\,[H]^2 = 0$$

$$\frac{d\,[C_2\,H_5]}{dt} = k_2\,[H]\,[C_2H_4] - k_3[C_2H_5]\,[H_2] = 0$$

and $2k_1\,[Hg]\,[H_2] - 2k_4\,[H]^2 = 0$

Solving for [H]

$$[H] = \left(\frac{k_1\,[Hg]\,[H_2]}{k_4}\right)^{1/2}$$

Substitute for [H] in equation (6) and solve for C_2H_5 to obtain

$$[C_2H_5] = \frac{k_2[C_2\,H_4]}{k_3\,[H_2]} \quad \left(\frac{k_1\,[Hg]\,[H_2]}{k_4}\right)^{1/2}$$

From equation (5)

$$\frac{d\,[C_2\,H_6]}{dt} = k_3\,\frac{k_2[C_2\,H_4]}{k_3\,[H_2]}\left(\frac{k_1\,[Hg]\,[H_2]}{k_4}\right)^{1/2} \quad [H_2]$$

$$= k_2\,[C_2H_4]\left(\frac{k_1\,[Hg]\,[H_2]}{k_4}\right)^{1/2}$$

Solution 2 : $\log \text{rate} = \log k + n \log C_i$

where k = rate constant

n = constant, called order of reaction (a and b in this case)

C_i = concentration of substance i

using equation (1), the order for both susbtances A and B can be determined. Using data from the first and third columns of table 1 and substituting the respective values into equation (1), gives

$$\log (0.05) = \log k_A + a \log 1$$

and

$$\log (0.20) = \log k_A + a \log 2,$$

keeping ($C_{B,0}$) fixed.

If the first equation is subtracted from the second, then

$$\log (0.20) - \log (0.05) = a \log 2 - a \log 1$$

$$\log (0.020/0.05)\ a\ (\log (2/1))$$

$$0.602 = a\ (0.301)$$

$$\therefore \quad a = \frac{0.602}{0.301} = 2$$

Doing the same for substance B, using the first and second columns and keeping $C_{A,0}$ fixed, gives

$$\log (0.05) = \log k_B + b \log 1$$

and

$$\log (0.10) = \log k_B + b \log 2$$

If the first equation is subtracted from the second, then

$$\log (0.10) - \log (0.05) = b \log 2 - b \log 1$$

$$\therefore \quad \log \frac{0.10}{0.05} = b \log (2/1)$$

$$(0.301) = b\ (0.301)$$

$$\therefore b = \frac{0.301}{0.301} = 1$$

Therefore, rate = $kC_A^2\ C_B$

Soloving for k, yields $\quad k = \dfrac{\text{rate}}{C_A^2 C_B}$

Using values of C_A, C_B and rate from table 1

$$k = \frac{0.05}{(1^2)(1)} = \frac{0.10}{(1^2)(2)} = \frac{0.20}{(2^2)(1)} = \frac{0.40}{(2^2)(2)}$$

from which $k = 0.05\ M^{-2}s^{-1}$ (no matter which column is chosen).

Solution 3: From the chemical reaction, the rate of change of the respective ions can be represented as

$$= \frac{d}{dt}(C_{CrO_4^{2-}}) = \frac{d}{dt}(CH+) = -2\frac{d}{dt}\left(C_{Cr_2O_7^{2-}}\right)$$

This is because both of the product ions are being formed twice as fast as the dichromate ion is reacting. For the water,

$$-\frac{d}{dt}(C_{H_2O}) = \frac{d}{dt} - \left(C_{Cr_2O_7^{2-}}\right)$$

because the water is reacting in a 1:1 ratio to $Cr_2O_7^{2-}$.

The complete rate equation is the second-order equation. Therefore,

$$\text{rate} = K\, C_{H_2O}\, C_{Cr_2O_7^{2-}}$$

K = rate constant

C = concentration

If the concentration of water remains constant, the revised rate equation will be the pseudo-first-order equation. Thus,

$$\text{rate} = K'\, C_{Cr_2O_7^{2-}}$$

where $K' = K\, C_{H_2O}$

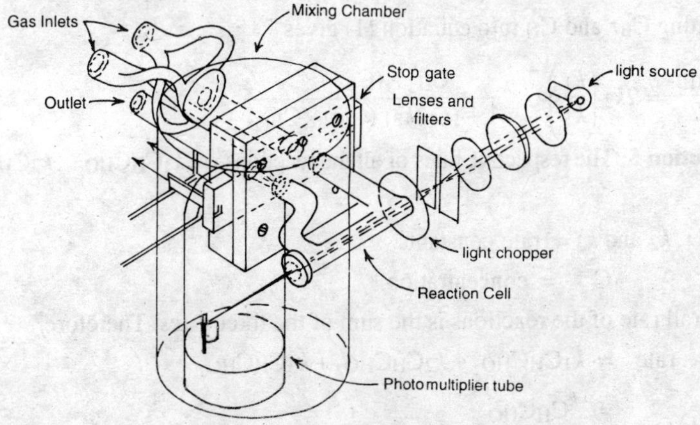

Apparatus of Johnston and Yost for study of a rapid gas reaction : isometric projection of mnixing chamber, stop gate, and 2 mm diameter reaction cell, with schematic drawing of the light source, filters, and lenses.

Solution 4 : From the second equation, $\frac{dC_{HBr}}{dt} = k_2 C_{Br} C_{H_2}$.

The third equation gives $\frac{dC_{HBr}}{dt} = k_3 C_H C_{Br2}$ and the fourh gives - $\frac{dC_{HBr}}{dt}$ $= k_4\ C_H C_{HBr}$.

Combining these three equations together yield

$$= \frac{dC_{HBr}}{dt} = k_2 C_{Br} C_{H_2} + k_3 C_{Br2} - k_4 C_{HBr} \quad (1)$$

Similarly,

$$\frac{dC_H}{dt} = 0 = k_2 C_{Br} C_{H_2} - k_3 C_H C_{Br_2} - k_4 C_H C_{HBr} \quad (2)$$

and

$$\frac{dC_{Br}}{dt} = 0 = 2k_1 C_{Br_2} - k_2 C_{Br H_2} + k_3 C_H C_{Br_2} + k_4 C_H C_{HBr}$$

$$- 2k_5 C_{Br}^2 \quad (3)$$

Solving equations (2) and (3) simultaneously yields

$$C_{Br} = \left(\frac{k_1}{k_5}\right)^{1/2} (C_{Br_2})^{1/2}$$

and

$$C_H = k_2 \left(\frac{k_1}{k_5}\right)^{1/2} \frac{(C_{Br_2})^{1/2}\ C_{H_2}}{k_3\ C_{Br_2} + k_4 C_{HBr}}$$

Substituting C_{Br} and C_H into equation (1) gives

$$\frac{dC_{HBr}}{dt} = 2k_2 \left(\frac{k_1}{k_5}\right)^{1/2} \frac{C_{H_2}\ (C_{Br_2})^{1/2}}{1 + (k_4/k_3)\ (C_{BHr}/C_{Br_2})}$$

Solution 5: The respective rates of all the reactions are $k_1 C_H C_{HO_2}$, , $k_2 C_H C_{HO_2}$ where

k_1, k_2 and k_3 = rate constants

C = concentration.

The overall rate of the reactions is the sum of the three rates. Therefore,

$$\text{rate} = k_1 C_H C_{HO_2} + k_2 C_H C_{HO_2} + k_3 C_H C_{HO_2}$$

$$= {}^k C_H C_{HO_2}$$

where $k = k_1 + k_2 + k_3$.

The given ratio of the rate constants implies that the products of the first reaction will be formed from 62% of the reactants will form the products of the third reaction.

Since the stiochiometry of the product of the second equation is 2, the ratio of the products will be

$C_{H_2} = C_{O_2}/C_{OH}/C_{H_2O} = C_O$

62% / 2 × 27% / 11%

or 0.62 / 0 0.54 / 0.11

Solution 6 : $= \frac{dC_C}{dt}\ k_1C_AC_B - k_{-1}C_C$ from the first reaction and $- \frac{dC_C}{dt} =$ $k_2C_CC_B$ from the second reaction. If the latter is subtracted from the former, then

$$dC_C/dt + \frac{dC_C}{dt} = k_1C_AC_B - k_{-1}C_c - k_2C_CC_B$$

Therefore, $0 = k_1C_AC_B - k_{-1}C_C - k_2C_cC_B$ (1)

Recall that dc_c/dt is assumed to be 0

Therefore,

$$dc_c/dt = 0 = k_1C_AC_B - k_{-1}C_c - k_2C_cC_B$$

since equation (1) is also equal to zero.

Solving for C_C,

$$k_{-1}C_c + k_2\ C_CC_B = k_1C_AC_B$$

$$C_C\ (k_{-1} + k_2C_B) = k_1C_AB_B$$

$$\therefore\quad C_C = \frac{k_1C_A\ C_B}{k_{-1} + k_2\ C_B} \tag{2}$$

From the two reactions,

$$-dC_A/dt = k_1C_AC_B - k_{-1}C_C \tag{3}$$

$$-dC_B/dt = k_1C_AC_B - k_{-1}C_C + k_2C_BC_C \tag{4}$$

$$\frac{dC_D}{dt} = k_2C_BC_C \tag{5}$$

Since the final answers are supposed to be in terms of C_A, C_B and C_D only, we can now substitue equation (2) into equations (3), (4) and (5) in order to eliminate C_C.

Therefore,

$$\frac{-dC_A}{dt} = k_1C_AC_B - \frac{k_{-1}\,k_1\,C_A\,C_B}{k_{-1} + k_2\,C_B}$$

$$= \frac{k_1\,C_A\,C_B\,(k_{-1} + k_2\,C_B) - k_{-1}\,k_1\,C_A\,C_B}{k_{-1} + k_2\,C_B}$$

$$= \frac{k_1 C_A C_B k_{-1} + k_1 k_2 C_A C_B^2 - k_{-1} k_1 C_A C_B}{k_{-1} + k_2 C_B}$$

$$= \frac{k_1 k_2 C_A C_B^2}{k_{-1} + k_2 C_B}$$

$$\frac{-dC_B}{dt} = k_1 C_A C_B - \frac{k_{-1} k_1 C_A C_B}{k_{-1} + k_2 C_B} + \frac{k_2 C_B k_1 C_A C_B}{k_{-1} + k_2 C_B}$$

Combining them yields

$$-dC_B/dt = \frac{2k_1 k_2 C_A C_B^2}{k_{-1} + k_2 C_B}$$

Finally,

$$dC_D/dt = \frac{k_2 C_B k_1 C_A C_B}{k_{-1} + k_2 C_B} = \frac{k_1 k_2 C_A C_B^2}{k_{-1} + k_2 C_B}$$

Solution 7 : (a) The rate law for reaction (3) [See Problem] is

$$\frac{d[NO_2]}{dt} = 2k_3 [N_2O_2][O_2] \qquad (1)$$

For reactions (1, 2, 3) the rate law becomes

$$\frac{d[N_2O_2]}{dt} = k_1 [NO]^2 - k_2 [N_2O_2] - k_3 [N_2O_2][O_2] = 0$$

Solving for $[N_2O_2]$ gives

$$[N_2O_2] = \frac{k_1 [NO]^2}{k_2 + k_3 [O_2]}$$

Substitute for $[N_2O_2]$ in equation (1) to obtain

$$\frac{d[NO_2]}{dt} = \frac{2k_1 k_3 [NO]^2 [O_2]}{k_1 + k_3 [O_2]}$$

(b) Assume a reversible reaction.

$$2NO \underset{k_2}{\overset{k_1}{\rightleftharpoons}} N_2O_2 \text{ or } K_{1,2} = \frac{[N_2O_2]}{[NO]^2} \qquad (2)$$

and

$$N_2O_2 + O_2 \xrightarrow{k_3} 2NO_2$$

But from equation (1)

$$= \frac{d[NO_2]}{dt} = 2k_3 [N_2O_2][O_2]$$

From equation (5), $[N_2O_2] = K_{1,2} [NO]^2$

$$\therefore \frac{d[NO_2]}{dt} = 2k_3 K_{1,2} [NO]^2 [O_2]$$

$$= k [NO]^2 [O_2]$$

where $k = 2k_3 K_{1,2}$ = third order rate constant

or $\ln k = \ln 2 + \ln k_3 + \ln K_{1,2}$

$$\frac{d \ln k}{dT} = \frac{d \ln k_3}{dT} + \frac{d \ln k_{1,2}}{dT}$$

and $Ea = E_3 + (E_1 - E_2) = 82kJ + (82 - 205)\ kJ = -41kJ$

Solution 8 :From kinetic theory of molecular collisions, the number of collisions, z of identical molecules per unit volume per unit time is given by

$$Z = 2\left(\frac{\pi RT}{M}\right)^{1/2} \sigma^2 n^2$$

where σ is the collision diameter for the molecules, n is the number of molecules per unit volume

$$n = \frac{\text{number of molecules}}{\text{volume}}$$

$$= \frac{\text{number of moles x number of molecules/mole}}{\text{volume}}$$

$$= \frac{mN_0}{V}$$

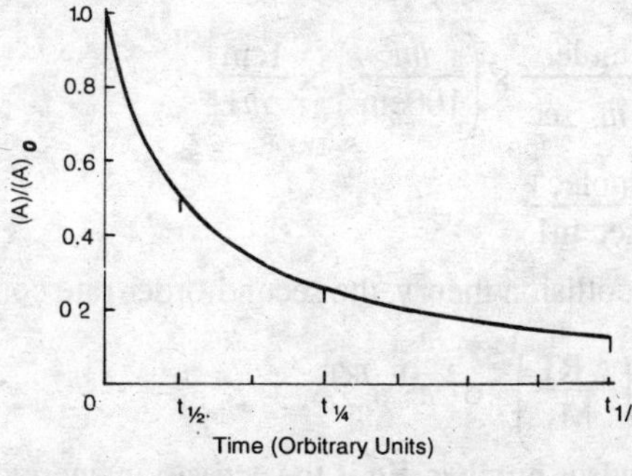

The second-order rate law. $(A)/(A)_1$ *as a funtion of time, showing the neccessive doubling of* $t_{1/2}$

where m is the number of moles, N_0 is Avogadro's number, and V is measured in cubic meters. The ideal gas law give

$$PV = mRT \text{ or } = \frac{P}{RT} = \frac{m}{V},$$

but $n = \frac{m}{V} N_0$ and therefore

$$n = \frac{N_0 P}{RT}$$

$$= \frac{(6.02\,(10^{23})\text{ mole c/mole}) \times (760\text{ torr}) \times (1\text{ atm}/760\text{ torr})}{(0.0821\text{ latm° K}^{-1}\text{ mole}^{-1})}$$

$$= 9.17\,(10^{21})\,\frac{\text{mole c}}{1} \times 1000\,\frac{l}{m^3} = 9.17\,(10^{24})\,\frac{\text{mole c}}{\text{m}^3}$$

and M is the molecular weight of acetaldehyde molecule.

From equation (1),

$$Z = 2 \times \left(\frac{\pi \times (8.314\text{J/°K }mole) \times (800\text{ °K})}{(4.4(10^{-2})\text{ kg/mole})}\right)^{1/2}$$

$$\times (5(10^{-10})m)^2 \times \left((9.17\,(10^{24})\,\frac{\text{molec}}{\text{m}^3}\right)^2$$

$$= 2 \times (4.747\,(10^5)\text{ J}/kg)^{1/2} \times 2.5\,(10^{-19})m^2 \times 8.39\,(10^{49})\text{ molec}^2/m^6$$

$$= 2 \times \left(\frac{4.747\,(10^5)\,kg\,m^2/\text{sec}^2}{kg}\right)^{1/2} \times (2.09\,(10^{31})\text{ molec}^2/m^4)$$

$$= 2 \times (688.96\,m/\text{sec}) \times 2.09\,(10^{31})\text{ molec}^2/m^4$$

$$= 2.88\,(10^{34})\,\frac{\text{molec}^2}{m^3\text{ sec}} \times \left(\frac{lm}{100\text{cm}}\right)^3 \times \frac{1\text{cm}^3}{m1}$$

$$= 2.88\,(10^{28})\,\frac{\text{molec}^2}{\text{sec ml}}$$

(b) Also from the gas collision theory, the second-order rate constant is given by

$$k = 2 \times 10^3\,N_A\,P\left(\frac{\pi RT}{M}\right)^{1/2} \sigma^2\,e^{-E_a/RT} \qquad (2)$$

where N_A = the Avogadros number, Ea is the activation energy and P is the steric factor, in this case taken as 1.

From equation (2),

$$k = (2 \times 10^3\,lm^{-3})\,(6.02 \times 10^{23}\text{ mole}^{-1})$$

$$\left[\frac{\pi\,(8.31\text{ J }K^{-1}\text{ mole}^{-1})\,(800°\text{K})}{4.4 \times 10^{-2}\,kg\text{ mole}^{-1}}\right]^{1/2}$$

$x\,(5 \times 10^{-10}\mu)^2$ [e (–45500 cal mole^{-1})/(1.987 cal °K^{-1} mole^{-1})

x (800 °K))] = $(2 \times 10^3)\,(6.02 \times 10^{23})\,(6.89 \times 10^2)\,(9.28 \times 10^{-32})$

= 0.077 liter mole $^{-1}$ sec $_{-1}$

Solution 9 : The relationship between the half-life and the initial concentration of a reaction is given by

$$\log t_{1/2} = \log \left[\frac{2^{n-1} - 1}{(n-1)\,k}\right] - (n - 1) \log C_{i,0} \qquad (1)$$

where $t_{1/2}$ = half-life period

$C_{i,0}$ = initial concentration

n-1 = slope of the plot log $t_{1/2}$ against log $C_{i,0}$

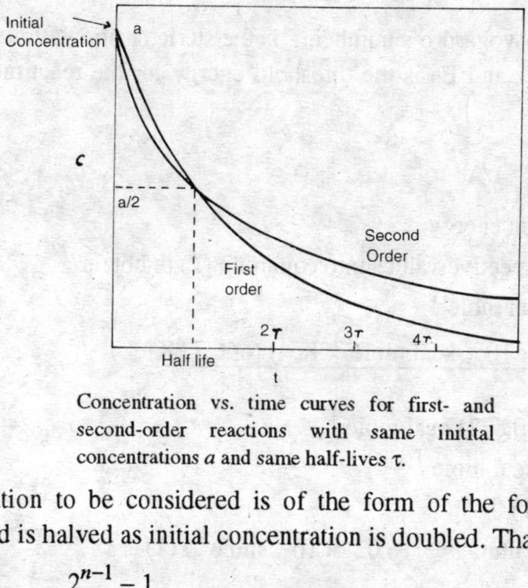

Concentration vs. time curves for first- and second-order reactions with same initial concentrations a and same half-lives τ.

Another equation to be considered is of the form of the former, except that half-life period is halved as initial concentration is doubled. That is

$$\log \left(\frac{t_{1/2}}{2}\right) = \log \frac{2^{n-1} - 1}{(n-1)\,k} - (n\text{-}1) \log 2C_{i,0} \qquad (2)$$

Subtracting equation (2) from equation (1) yields

$$\log t_{1/2} - \log \frac{t_{1/2}}{2} = -(n\text{-}1) \log C_{i,0} + (n\text{-}1) \log 2C_{i,0}$$

$$\log \left(t_{1/2} \Big/ \frac{t_{1/2}}{2}\right) = (n\text{-}1) \log 2C_{i,0} - (n\text{-}1) \log C_{i,0}$$

$$\log 2 = (n - 1)\,[\log 2C_{i,0} - \log C_{i,0})$$

$$\log 2 = (n\text{-}1) \log \frac{2C_{i,0}}{C_{i,0}} = (n\text{-}1) \log 2$$

$\therefore \quad n\text{-}1 = \log 2/\log 2 = 1$

or $\quad n = 1+1 = 2$

Solution 10: The collision theory postulates that in order for molecules to interact, they must approach each other so closely that they can be said to be in collision. The speed of the reaction is equal to the number of collisions per second times the fraction of the collisions that are effective in producing chemical change.

Thus, according to this gas-collision theory, the second-order rate constant should be giveb by

$$k = 2 \times 10^3\, N_A P \left(\frac{\pi\, PT}{M}\right)^{1/2} \sigma^2\, e(E_o/RT) \qquad (1)$$

where N_A = the Avogardro's number, P is the steric or orientation factor, σ is the collision diameter and E_0 is the threshold energy for the reaction, given by the expression

$$E_0 = E_a - \frac{RT}{2} \qquad (2)$$

E_a is the activation energy

Substitute the respective values into equaition (2) to obtain

$$E_0 = 45.6\ k\text{cal mole}^{-1}$$

$$- \frac{(1.987 \times 10^{-3}\ \text{kcal mole}^{-1}\,{}^\circ K^{-1})\ (666.7\ {}^\circ K)}{2}$$

$$= (45.6 - 0.622)\ \text{kcal mole}^{-1}$$

$$= 44.94\ \text{kcal mole}^{-1}$$

From equation (1),

$$k = 2 \times (10^3\ \text{liters}\ m^{-3})\ (6.02\ \times 10^{23}\ \text{mole}^{-1})\ (1)$$

$$x \left[\frac{\pi\ (8.31\ \frac{J\,K^{-1}}{\text{mole}})\ (666.7\ {}^\circ K)}{127.9 \times 10^{-3}\ \text{kg mole}^{-1}}\right]^{1/2}$$

$$x \left[3.5\mathring{A} \left(\frac{10^{-10} m}{1\mathring{A}}\right)\right]^2 e^{-(44.94\text{kcal mole}^{-1})/(1.987\text{cal}^\circ K^{-1}\text{mole}^{-1})}$$

$$x\ (666.7^\circ K)$$

$$= 1.0 \times 10^{-4}\ M^{-1} S^{-1}$$

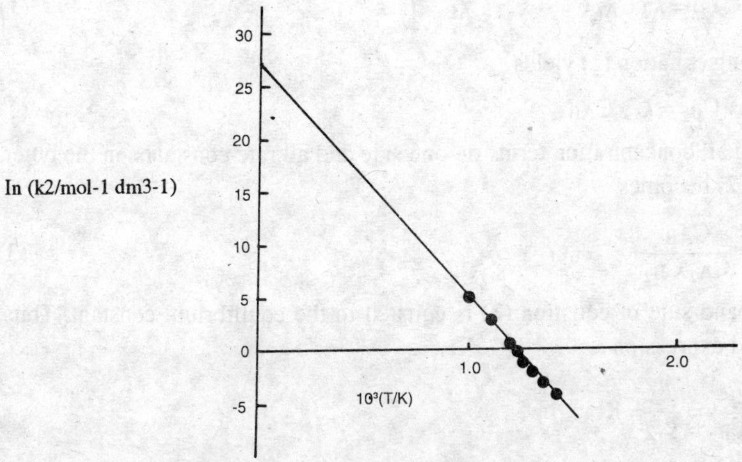

Solution 11 : The variation of temperature with reaction rate constants is usually expressed by the Arrhenius equation,

k_2 = A*e* - E*a*/RT (1)

where A is the pre-exponential factor and E*a* is the activation energy.

Equation (1) can be written in a logarithmic form as follows :

In k_2 = - + In A(2)

From equation (2), it can be seen that a straight line will be obtained if the logarithm of the rate constant is plotted against the reciprocal of the absolute temperature.

From the plot, the sope is determined to be -2.207×10^4 °K)]

Thus $E_a = -[(8.314 J\ K^{-1}\ mole^{-1})\ (-2.207 \times 10^4\ °K)$

$= 1.84 \times 10^5$ J $mole^{-1}$

= 184 *k*J $mole^{-1}$

The intercept on the plot has a value of 26.95 and thus

$A = 5.06 \times 10^{11}$.

Solution 12: Opposing reactions proceed at equal rates when at equilibrium. Therefore, there is a net rate of change of zero for each component. The thermodynamic equilibrium constant, K can be obtained by first collecting all concentration terms on one side and all rate constants on the other side. Finally, the latter side is equated to the equilibrium constant.

From the reaction under consideration, the expression for component A at equilibrium i s

$$-\frac{d}{dt}(C_{A_2}) = k_2\, C_{A_1}\, C_{B_2} - k_{-2}\, C_{AB}^2 = 0 \qquad (1)$$

Rearranging equation (1) yields

$$k_2\, C_{A_1}\, C_{B_2} = k_{-2}\, C_{AB}^2 \qquad (2)$$

Collecting all concentration terms on one side and all rate constants on the other, equation (2) becomes

$$\frac{k_2}{k_{-2}} = \frac{C_{AB}^2}{C_{A_2}\, C_{B_2}} \qquad (3)$$

The left hand side of equation (3) is equated to the equilibrium constant. Thus, the desired expression is

$$\frac{C_{AB}^2}{C_{A_2}\, C_{B_2}} = \frac{k_2}{k_{-2}} = K.$$

Observe the K is the ratio of product concentrations to reactant concentrations and not vice versa.

Solution 13: Write a general rate law since the order of the reaction was not specified. That is,

$$\frac{dx}{dt} = k\,(a-x)^n \qquad (1)$$

where n = the order of the reaction, a is the initial concentration of the reactant and $(a-x)$ is the concentration of the reactant remaining after time, t.

Rearranging equation (1) yields

$$\frac{dx}{(a-x)^n} = k\,dt \qquad (1a)$$

Integrating equation (1a) gives

$$\frac{1}{(n-1)\,(a-x)^{n-1}} = kt + C \qquad (2)$$

where C = the integrating constant. To evaluate C, use the condition that at $t = 0$, $x = 0$. Therefore, equation (2) gives

$$\frac{1}{(n-1)\,a^{n-1}} = C$$

Rewrite equation (2), substituting for C, to obtain

$$\frac{1}{(n-1)\,(a-x)^{n-1}} = kt + \frac{1}{(n-1)\,a^{n-1}}$$

or

$$k = \frac{1}{t\,(n-1)\,(a-x)^{n-1}} - \frac{1}{t\,(n-1)\,a^{n-1}}$$

$$= \frac{1}{t\,(n-1)}\left[\frac{1}{(a-x)^{n-1}} - \frac{1}{a^{n-1}}\right]$$

The reaction is (0.20) complete at the time intervals given. Therefore, when $t = t_{1/5}$, $(a\text{-}x) = 0.80a$

Write the Arrhenius equation for the rate constant as

$k = A \exp(-E_a/RT)$ (3)

where A = the constant of integration or the pre-exponential factor and Ea is the activation energy. Substitute the expression for k in equation (3) to obtain

$$A^{-Ea/RT} = \frac{1}{t\,(n-1)}\left[\frac{1}{(0.80a)^{n-1}} - \frac{1}{a^{n-1}}\right]$$

or

$$e^{-E_a/RT} = \frac{1}{t}\;\frac{1}{A(n-1)}\left[\frac{1}{(0.80a)^{n-1}} - \frac{1}{a^{n-1}}\right] \tag{4}$$

Let

$$\frac{1}{A(n-1)}\left[\frac{1}{(0.80a)^{n-1}} - \frac{1}{a^{n-1}}\right] = \text{constant} = L$$

Therefore equation (4) changes to

$$e^{-E_a/RT} = \frac{1}{t}\,.\,L$$

Taking the ln of both sides yields

$ln\, e^{-E_a/RT} = ln\, L - ln\, t$

or

$$-\frac{E_a}{RT} = ln\, L - ln\, t$$

At 300 °K the reaction is 20% complete in 12.6 min.

Thus,

$$+\frac{E_a}{300R} = ln\, 12.6 - ln\, L \tag{5}$$

At 340 °K the reaction is 20% complete in 3.20 min

and

$$+\frac{E_a}{340R} = ln\, 3.20 - ln\, L \tag{6}$$

Subtract equation (6) from eqation (5) to get

$$0.04717\ Ea = ln\frac{12.6}{3.20}$$

$$= 1.371$$

$$\therefore \quad Ea = \frac{1.371}{0.04717}$$

$$= 29.06\ kJ$$

Solution 14 : Plot ln (P/Po) vs time, t. Po is the vapour pressure of azomethane at 600 °K. Since the plot gives a straight line, the reaction is first order, with a slope of $-k^{1.}$ From the graph, the slope is - 3.6 10^{-4}. Thus the rate constant $k_1 = 3.6 \times 10^{-4}$ s.

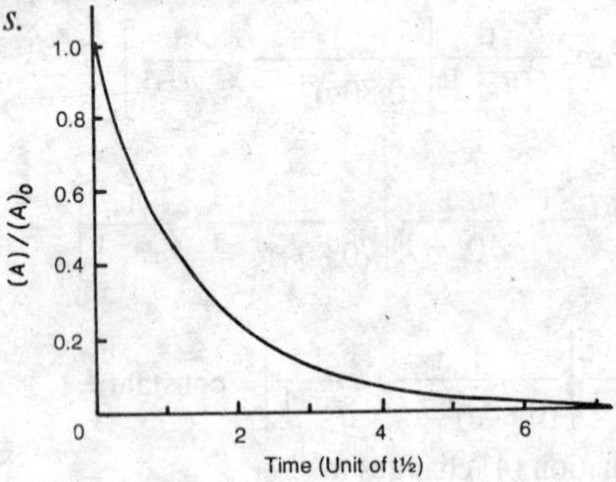

The first-order rate law, (A)/(A)o as a function of time, showing the constancy of $t_{1/2}$

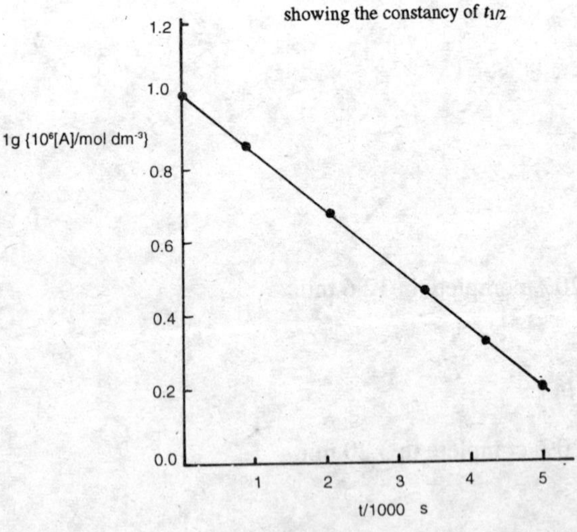

Determination of the rate coefficient for a first-order reaction.

Solution 15 : For a first order reaction the rate equation is written as

$$-\frac{d(A)}{dt} = k\,(A)$$

or $$-\frac{d(A)}{(A)} = kdt \qquad (1)$$

Let $(A)_1$ = concentration of A at $t = t_1$

and $(A)_2$ = concentration of A at $t = t_2$

Integrating equation (1) using the above limits yields

$$\int_{(A)_1}^{(A)_2} \frac{d(a)}{(A)} = k \int_{t_1}^{t_2} dt$$

$$In\,\frac{(A)_1}{(A)_2} = k\,(t_2 - t_1) \qquad (2)$$

Assume that $t_1 = 0$ and that the initiated concentration of A = $(A)_0$, equation (2) becomes

$$In\,\frac{(A)_0}{(A)} = kt.$$

Taking exponents of both sides gives

$$e^{In\,\frac{(A)_0}{(A)}} = e^{kt}.$$

and

$$\frac{(A)_0}{(A)} = e^{kt}$$

Solving for (A) yields

$$(A) = (A)_0\, e^{-kt}$$

Taking *ln* of both sides gives

$$ln(A) = ln(A)_0\, e^{-kt}$$

$$= -kt + ln(A)_0$$

or

$$\log\,(A) = \frac{-kt}{2.303} + \log\,(A)_0 \qquad (3)$$

(*a*) In this problem $(A)_0 = 100\%$ and $(A) = 25\%$ when $t = 1$ hr.

Therefore, solve equation (3) for *k* as follows

$$\log 0.25 = \frac{-k\,(3600\text{ sec})}{2.303} + \log 1$$

$$- 0.602 = -1563.18\, k \text{ sec} + 0$$

and $k = 3.85 \times 10^{-4}\ \text{sec}^{-1}$

Now, use this value of k to solve eqaution (3) for (A) when t = 2 hrs.

$$\log A = \frac{(-3.85 \times 10^{-4}\ \text{sec}^{-1})\ (7200\ \text{sec})}{2.303} + 0$$

$$= -1.2036$$

and $A = 6.26 \times 10^{-2}$

$$= 6.26\%$$

(b) Here use the second order equation to find K.

The second order equation is written as

$$kt = \frac{1}{(A)} - \frac{1}{(A)_0} \qquad (4)$$

$$k(3600\ \text{sec}) = \frac{1}{0.25} - 1$$

$$k = \frac{3}{3600\ \text{sec}}$$

$$k = 8.33 \times 10^{-4}\ \text{sec}^{-1}$$

Use this value of k to solve for A in equation (4). From equation (4),

$$(8.33 \times 10^{-4}\ \text{sec}^{-1})\ (7200\ \text{sec}) = \frac{1}{(A)} - 1$$

$$(A) = 1.429 \times 10^{-1}$$

$$= 14.3\%$$

(c) The reaction continues at constant rate. Since 75% reacts in 1 hr., the reaction will be complete before 2 hrs. Therefore the answer to this section is 0%.

Solution 16: One way to determine the order of the reaction is to plot log (∞ V) vs t. A straight line would be obtained if the reaction is first order. From the slope of the plot, the value of k can be calculated.

Another method is to use the equation

$$k = \frac{2.303}{t} \log \frac{V_\infty - V_0}{V_\infty - V}$$

For $t = 4$ min

$$k = \frac{2.303}{4\ \text{min}} \log \frac{15.98 - 1.71}{15.98 - 3.03}$$

$$= 0.0243\ \text{in}^{-1}$$

For t = 4 min to 5 = 50 min

k = 0.0243, 0.0241, 0.0236, 0.0231, 0.0232, 0.0233 and 0.0241.

Since k shows constancy then it can be said that the reaction is first order. And the value for k, taken as an average of all the values, is approximately 0.0237.

Solution 17: (a) In a given reaction, the total pressure is the sum of the partial pressures. That is

$$P = \Sigma P_i$$

where P = total pressure

P_i = partial pressure of component i

subscripi i = the species, which are C_5H_8, H_2 and C_5H_6 in this problem.

For this problem,

$$P = P_{C_5H_8} + P_{H_2} + P_{C_5H_6}$$

The stiochiometry of the reaction is such that

$$P_{H_2} = P_{C_5H_6} = P_{C_5H_8,0} - P_{C_5H_8}$$

where $P_{C_5H_8,0}$ = partial pressure of C_5H_8 at time zero

$$\therefore \quad P = P_{C_5H_8} + 2(P_{C_5H_8}0 - P_{C_5H_8})$$

Now

$$P = P_{C_5H_2} + 2(P_{C_5H_8})) - P_{C_5H_8} = 2P_{C_5H_8,0} - P_{C_5H_8} \quad (1)$$

Therefore differentiating (1) yields

$$\frac{dP}{d}t = -\frac{dP_{C_5H_8}}{dt} = k\,P_{C_5H_8} \quad (2)$$

but

$$P_{C_5H_8} = k\,C_{C_5H_8}\text{, so}$$

$$\frac{dP_{C_5H_8}}{dt} = k\frac{dC_{C_5H_8}}{dt}\text{, hence } \frac{dP}{dt} = -\frac{dP_{C_5H_8}}{dt} = -k\frac{dC_{C_5H_8}}{dt}$$

(b) Observe that the units of k are 1/time, where time is in seconds.

(c) From equation (2),

$$-\frac{dP_{C_5H_8}}{dt} = k\,P_{C_5H_8}$$

$$\therefore \quad \frac{dP}{dt} = k\,P_{C_5H_8} = k\,(2P_{C_5H_8,0} - P)$$

Rearranging and then integrating both sides, give

$$ln\left[\frac{P_{C_5H_8,0}}{(2P_{C_5H_8,0} - P)}\right] = kt$$

Solution 18: The reaction equation can be written as

$$C_3H_7Br + S_2O_{32-} \rightarrow C_3H_7S_2O_{3-} + Br^-$$

$$I_2 + 2S_2O_3^{2-} \times 2I^- + S_4O_{6-2}$$

Let the initial concentrations at $t = 0$ be "a" mole cm^{-3} of S_2O_{3-2} and "b" moles cm^{-3} of C_3H_7Br

But $$a = \frac{I_2 \text{ consumed at } t = 0}{\text{Volume } S_2O_3^{2-}}$$

$$= \frac{37.63 \times 0.02572\text{N} \times \dfrac{1 \text{ mole } I_2}{\text{eq transferred}}}{10.02}$$

$$= 0.09659 \text{ M}$$

$$b = \frac{\text{total } S_2O_3^{2-} \text{ consumed}}{\text{Volume } S_2O_3^{2-}}$$

$$= \frac{(37.63 - 22.24)0.02572\text{M}}{10.02}$$

$$= 0.03950 \text{ M}$$

After a time t, x mole cm^{-3} of $S_2O_3^{2-}$ and C_3H_7Br will have reacted forming x mole cm^{-3} of $C_3H_7S_2O_{3-}$ and of Br^-.

Therefore, the concentration of $S_2O_3^{2-}$ at time t

$$(a\text{-}x) = \frac{0.02572 \text{ V}_{I_2}}{10.02}$$

and the concentration of C_3H_7Br at time t

$$(b\text{-}x) = \frac{0.02572(\text{V}_{I_2} - 22.24)}{10.02}$$

and $(a\text{-}b) = 0.05709$ M

The integrated second order rate law is

$$\frac{1}{a-b} \, ln = \frac{b\,(a-x)}{a\,(b-x)} \, k_2 t$$

and

$$k_2 = \frac{1}{t\,(a-b)} \ln \frac{b\,(a-x)}{a\,(b-x)}$$

$$= \frac{1}{(.05709M)\,t} \ln \left[\frac{(0.0395M)\left(\dfrac{.02572\,V_{I_2}}{10.02}\right)}{(.09659M)\left(\dfrac{.02572\,(V_{I_2} - 22.24)}{(10.02)}\right)} \right]$$

$$= \frac{1}{(.05709M)\,t} \ln \left[\frac{(0.03950)(V_{I_2}}{(.09659)\,V_{I_2} - 22.24)} \right] ln$$

The values of k_2 in liters mole^{-1} sec^{-1} evaluated at the different concentration levels are 0.001657, 0.001643, 0.001643, 0.0016149, 0.001618 and 0.001618 respectively. This confirms that the reaction is second order because of the consistency of k values.

The rate constant can be assumed to be an average of the above values of k_2. It is equal to .001640 l mole^{-1} sec^{-1}.

Solution 19: The rate equation for a number of reactions in a complex mechanism is written as a sum of the rate equations for the simple reactions making up the complex mechanism.

So, for the substance A,

$$= \frac{dC_A}{dt} k_1 C_A C_B - k_{-1} C_C$$

Note that the rate equation for A contains two terms (in spite of the fact that it appears only once). This is because it reacts in and is formed by this reaction, as indicated by the arrows going in opposite directions.

Using the same criterion, substance B should contain three terms because it reacts in and is formed by the first reaction. In addition to these, it reacts in the second equation.

Therefore,

$$-\frac{dC_B}{dt} = k_1 C_A C_B - k_{-1} C_C + k_2 C_C C_B$$

Similarly,

The substance D appears only once in the second equation.

Since it is being formed, its rate equation is represented by

$$\frac{dC_D}{dt} = k_2 C_C C_B$$

Solution 20: The rate equation for this mechanism is written as the sum of the rate equations for the simple reactions that make up the complex mechanism. Therefore, the rate equation for the enzyme subtrate complex X, will contain four terms. That is

$$\frac{dC_X}{dt} = 0 = k_1 C_E C_S - k_{-1} C_X - k_2 C_X + k_{-2} C_E C_P \qquad (1)$$

where C = concentration

k_1, k_{-1}, k_2, k_{-2} are rate constants

In the same manner,

$$\frac{dC_P}{dt} = k_2 C_X - k_{-2} C_E C_P \qquad (2)$$

Now, let $C_E + C_X = C_{E,0}$, $V_X = k_2 C_{E,0}$,

$$V_P = k_{-1} C_{E,0}, \; K_S = \frac{k_{-1} + k_2}{k_1} \text{ and } K_P = \frac{k_{-1} + k_2}{k_{-2}}$$

Consequently, eliminating C_X and C_E gives

$$\frac{dC_P}{dt} = \frac{(V_S/K_S)C_S - (V_P/K_P)\,C_P}{1 + (C_S/K_S) + (C_P/K_P)} \qquad (3)$$

Or instead of using these substitutions, it might be easier to solve for C_X in equation (1) and substitute its expression into equation (2). That is

$$0 = k_1 C_E C_S - k_{-1} C_X - k_2 C_X + k_{-2} C_E C_P$$

$$\therefore \quad k_{-1} C_X + k_2 C_X = k_1 C_E C_S + k_{-2} C_E C_P$$

$$C_X = \frac{k_1 C_E C_C + k_{-2} C_E C_P}{k_{-1} + k_2}$$

Substituting this into equation (2) gives

$$\frac{dC_P}{dt} = k_2 \left[\frac{k_1 C_E C_S + k_{-2} C_E C_P}{k_{-1} + k_2}\right] - k_{-2} C_E C_P \qquad (4)$$

Now what is done is to get the same denominator for both terms, so $k_{-2} C_E C_P$ is multiplied by

$$\frac{k_{-1} + k_2}{k_{-1} + k_2}$$

Hence,

$$\frac{dC_P}{dt} = k_2 \left[\frac{k_1 C_E C_S + k_{-2} C_E C_P}{k_{-1} + k_2}\right] - k_{-2} C_E C_P \left[\frac{k_{-1} + k_2}{k_{-1} + k_2}\right]$$

$$= \frac{k_2 k_1 C_E C_S}{k_{-1} + k_2} + \frac{k_2 k_{-2} C_E C_P}{k_{-1} + k_2} - \frac{k_{-2} k_{-1} C_E C_P}{k_{-1} + k_2} + \frac{k_2 k_{-2} C_E C_P}{k_{-1} + k_2}$$

Let $$K_S = \frac{k_{-1} + k_2}{k_{-1}} \text{ and } K_P = \frac{k_{-1} + k_2}{k_{-2}}$$

Therefore,

$$\frac{dC_P}{dt} = \frac{k_2}{K_S} C_E C_S + \frac{k_2 C_E C_P}{k_P} - \frac{k_{-1} C_E C_P}{k_P} - \frac{k_{-2} C_E C_P}{k_P}$$

Now $\quad C_{E,0} = C_E + C_X'$ So $C_E = C_{E,0} - C_X$

Hence,

$$\frac{dC_P}{dt} = \frac{k_2 C_S [C_{E,0} - C_X]}{K_S} + \frac{k_2 C_P}{K_P}[C_{E,0} - C_X] - \frac{k_{-1}[C_{E,0} - C_X]}{K_P} - \frac{k_2 C_P}{K_P}$$

$$\times [C_{E,0} - C_X]$$

$$= \frac{k_2 C_S C_{E,0}}{K_S} - \frac{k_2 C_S C_X}{K_S} + \frac{k_2 C_P C_{E,0}}{K_P} - \frac{k_2 C_S C_X}{K_P} - \frac{k_{-1} C_P C_{E,0}}{K_P}$$

$$+ \frac{k_{-1} C_S C_X}{K_P} - \frac{k_{-2} C_P C_{E,0}}{K_P} + \frac{k_2 C_P C_X}{K_P}$$

Let $k_{-1} C_{E,0} = V_P$ and $k_2 C_{E,0} = V_S$. Then

$$\frac{dC_P}{dt} = \frac{V_S C_S}{K_S} - \frac{k_2 C_S C_X}{K_S} + \frac{V_S C_P}{K_P} - \frac{k_2 C_P C_X}{K_P} - \frac{V_P C_P}{K_P} + \frac{k_{-1} C_P C_X}{K_P}$$

$$- \frac{C_P V_S}{K_P} + \frac{k_2 C_P C_X}{K_P}$$

$$= \frac{V_S}{K_S} C_S - \frac{V_P}{K_P} C_P + \frac{k_{-1} C_P C_X}{K_P} - \frac{k_2 C_S C_X}{K_S}$$

Now, substitute in the value of C_X which was determined earlier after simplifying, the same results will be found as in equation (3)

If the measurements of the reaction rate are made during the early stages of the reaction, C_P will be nearly zero and the rate equation simplifies to give the Michaelis-Menten equation :

$$\frac{dC_P}{dt} = \frac{V_S/K_S)C_S}{1 + (C_S/K_S)} = \frac{V_S}{1 + (K_S/C_S)}$$

if $C_S < K_S$, the reaction rate is first-order with respect to C_S and if $C_S > K_S$, the reaction rate is zero-order with respect to CS. In both cases the reaction is dependent on V_S, which is first-order with respect to $C_{E,0} = C_E$. The Michaelis constant for the substrate, K_S, represents the value of C_S necessary to reduce the maximum reaction rate, V_S, by a factor of two. The rate constant for the product reaction, k_2, is known as the turnover constant. Casting the rate equation into the Lineweaver-Burk linear form gives

$$\left(\frac{dC_O}{dt}\right)^{-1} = \left(\frac{K_S}{V_S}\right) C_S^{-1} + V_S^{-1}$$

Thus a plot of $(dc_P/dt)^{-1}$ against C_S^{-1} will be linear, having an intercept of V_S^{-1} and a slope of K_S/V_S

Solution 21 : Be definition, if ψ is negative, then glycerol will be negatively adsorbed.

Assuming that this is a dilute solution, the γ term is essentially unity. Using the expression for the activity,

$$a_i = \gamma_i x_i, \text{ where}$$

γ_i = activity coefficient for substance *i*. Then $\gamma_2 = 1$, and $a_2 = x_2$. *Thus the* a_2 term can be approximated by x_2.

Therefore,

$$\psi = -\frac{X_2}{RT}\frac{dr}{dx_2} \quad (1)$$

where a_2 = activity

l = surface tension

x_2 = mole fraction (or composition)

R = universal gas constant

T = temperature

Since $\frac{dr}{dx_2}$ is positive, ψ will have a negative value. Consequently, glycerol will be negatively adsorbed.

Solution 22: A simple mechanism that describes an enzyme catalysis may be represented by

$$E + S \underset{k_{-1}}{\overset{k_1}{\rightleftarrows}} ES \; E \xrightarrow{k_2} + P$$

where E = enzymatic site

S = the substrate

ES = the enzyme-substrate complex

P = product of the reaction

If $k_{-1} >> k_2$, that is, if the rate of dissociation of the enzyme-substrate complex is much greater than the rate of its decomposition to form products, then the heat and entropy of formation of the complex can be obtained by using the Michaelis constant, K_M.

This is given by the expression

$$\frac{d \ \ln K_M}{dT} = \frac{\Delta H^\circ}{RT^2} \tag{1}$$

from which

$$\Delta G^\circ = -RT \ ln \ K_M \tag{2}$$

and $$\Delta S^\circ = - \frac{\Delta G^\circ - \Delta H^\circ}{T} \tag{3}$$

ΔH = heat of formation

ΔG = Gibbs free energy change

R = universal gas constant

T = temperature

ΔS = entropy of formation

The superscript in these equations refer to standard condition which for this problem can be taken to be 1 atm pressure and 15°C.

Before the above relations can be used, K_M must be determined by using the relation

$$\frac{V}{[S]} = \frac{V_M}{K_M + [S]}$$

or $$\frac{[S]}{V} = \frac{1}{V_M}[S] + \frac{K_M}{V_M}$$

where [S] = concentration of substrate, in this case luciferin

V = initial rate

V_M = maximum rate

The plot of $\frac{[S]}{V}$ against [S] should give a slope of $\frac{£}{V_M}$ and and an intercept of.

Then K_M is found.

Note that

$$K_M \ (\text{mols liter}^{-1}) = K_M \ (\text{ml stock}) \times \frac{\text{molarity stock}}{\text{Volume solution}}$$

The K_M values are 0.24 ml stock and 0.34 ml stock at 15°C and 22°C respectively. Converting these to mols liter^{-1}

$$K_M = \frac{0.24 \times 8.0 \times 10^{-5}}{20} = 0.096 \times 10^{-5} \text{ at } 15°X$$

and $$K_M = \frac{0.34 \times 8.0 \times 10^{-5}}{20} = 0.136 \times 10^{-5} \text{ at } 22°C$$

All values are tabulated in table 1.

Table 1:

[S]	(ml stock)	0.04	0.06	0.08	0.10	0.20	0.40	0.90
$V_{15°}$	(mV min-1)	15	19	23	26	44	56	76
$V_{22°}$	(mV min-1)	18	22	33	37	62	91	123
$\frac{[S]}{V15°}\times 10^3$		2.67	3.16	3.48	3.85	4.55	7.14	11.84
$\frac{[S]}{V22°}\times 10^3$		2.22	2.73	2.42	2.70	3.23	4.40	7.32

	15°	22°
V_M (mV min^{-1})	93	169
K_M (ml stock)	0.24	0.34
K_M (mols liter^{-1}) $\times 10^6$	0.96	1.36

From equation (1)

$$\Delta H° = \frac{RT_1 T_2}{T_2 - T_1} \ln \frac{K_M (T_2)}{K_M (T_1)}$$

$$= \frac{1.987 \times 288 \times 295}{295 - 288} \ln \frac{0.34}{0.24}$$

$$= 8400 \text{ cal/mol}$$

This is the amount of heat for the dissociation of the complex.

$\therefore \quad \Delta H°_{formation} = -8400 \text{ cal mol}^{-1}$.

Using equation (2)

$$\Delta G° = -1.987 \times 288 \ln 9.6 \times 10^{-7}$$

$$= 7930 \text{ cal/mole}$$

$\therefore \quad \Delta G°_{formation} = -7930 \text{ cal mol}^{-1}$

From equation (3)

$$\Delta S° = \frac{-7930 + 8400}{288} = 1.63 \text{ cal °K}^{-1} \text{ mol}^{-1}$$

$\Delta S°_{formation} = -1.63$ cal °K^{-1} mol^{-1}. The negative sign is due to the fact that upon association entropy decreases. The heat activation for the formation of product from the enzyme susbtrate complex can be determined from the effect of temperature on k_2 (the turnover number)

$$k_{2\,(15° C)} = \frac{V_M}{\text{enzyme concentration}}$$

$$= \frac{93mV\ min^{-1} \times 7.7 \times 10^{-12}\ mol\ mV^{-1}}{0.02\ liter \times 10^{-9}\ mol\ liter^{-1} \times 60\ sec\ min^{-1}}$$

$$= 0.597\ sec^{-1}$$

Similarly,

$$k_{2\ (22°C)} = \frac{169 \times 8.6 \times 10^{-12}}{0.02 \times 10^{-9} \times 60}$$

$$= 1.211\ sec^{-1}$$

$$\Delta H \text{ of activation} = Ea - RT \qquad (4)$$

for liquids and solids

where E_a = energy of activation

$T = (T_1 + T_2) / 2$

Approximating Ea from the k_2 values gives

$$E_a = \frac{RT_2\ T_1}{T_2 - T_1} ln \left(\frac{k(T_2)}{k(T_1)}\right)$$

$$= \frac{1.987 \times 288 \times 295}{295 - 288} ln \left(\frac{1.211}{0.597}\right)$$

$$= 17000\ cal\ mol^{-1}$$

Substitute the value of Ea into equation (4) to get

$$\Delta H \text{ of activation} = 17000 - 1.987 \times 291.5$$

$$= 16400\ cal\ mol^{-1}$$

Using the transition state theory,

$$\Delta S_{288}°K \text{ of activation} = \frac{\Delta H}{m} + R\ ln \left(\frac{k_2 h}{kT}\right)$$

where $h = 6.62 \times 10^{-27}$ erg sec = Planck's constant

$k = 1.38 \times 10^{-16} \left(\frac{erg}{°K}\right)$ = Boltzmann's constant

$$\therefore \quad \Delta S_{288}°K = \frac{16400}{288}$$

$$+ 1.987\ ln \left[\frac{0.597\ sec^{-1} \times 6.62 \times 10^{-27}\ erg\ sec}{1.38 \times 10^{-16} \frac{erg}{°K} \times 288°K}\right]$$

$$= -2.54\ cal\ deg^{-1}\ mol^{-1}$$

Solution 23: The rate equation for this system is

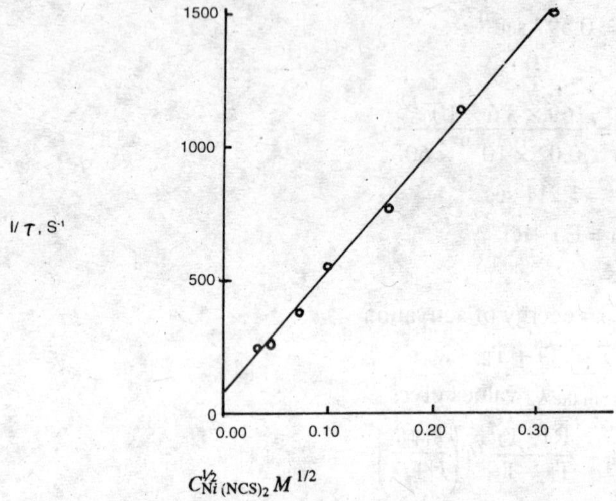

$$\frac{dc_{Ni(NCS)_2}}{dt} = k_f C_{Ni\,(NCS)+} C_{NCS-} - k_r C_{Ni\,(NCS)2}$$

At equilibrium, $k_f C_{Ni\,(NCS)+,e}\, C_{NCS-,e} - k_r C_{Ni(NCS)2,\,e} = 0$

For a small increase in, $C_{Ni(NCS)_2}$, $\Delta C_{Ni(NCS)_2}$

what is written is

$$C_{Ni(NCS)_2} = C_{Ni(NCS)_2\,,\,e} + \Delta C_{Ni(NCS)_2}$$

$$C_{NCS-} = C_{NCS-,\,e} - \Delta C_{Ni\,(NCS)_2}$$

$$C_{Ni\,(NCS)+} = C_{NCS+,\,e} - \Delta C_{Ni\,(NCS)_2}$$

The rate equation for the decay of $\Delta C_{Ni(NCS)_2}$ is

$$\frac{d\Delta C_{Ni\,(NCS)_2}}{dt} = kf\,(C_{Ni\,(NCS)_2,\,e} - \Delta C_{Ni\,(NCS)_2})\,(C_{(NCS)_2,\,e} + \Delta C_{Ni\,(NCS)_2})$$

$$-kr\,(C_{Ni\,(NCS)_2,\,e} + \Delta C_{Ni\,(NCS)_2})$$

$$= kr\,(C_{Ni\,(NCS)_2,\,e} + \Delta C_{Ni\,(NCS)_2})$$

$$= kf\,C_{Ni\,(NCS),\,+\,e}\;\; C_{NCS-,e} - C_{NCS-e}\,\Delta\,C_{Ni\,(NCS)_2} - C_{NI(NCS)+}$$

$$,e\,\Delta C_{Ni(NCS)_2} + \Delta C^2_{Ni(NCS_2}\,) - k_r\,(C_{Ni(NCS)_2,e} + \Delta C_{Ni(NCS)_2})$$

$$\therefore \qquad \frac{d\Delta C_{Ni\,(NCS)_2}}{dt} = -\,(k_f\,C_{Ni\,NCS)+,e} + k_f C_{NCS-,e} + k_r)$$

$$\times\,\Delta C_{Ni\,(NCS)_2} \qquad (1)$$

It is known that the displacement from equilibrium at time t is

$$\Delta C_i = \Delta C_{i,0}\, e^{-t/\tau} \qquad (2)$$

where $\Delta C_{i,0}$ = initital displacement from equilibrium. If equation (2) is differentiated then

$$\frac{d(\Delta C_i)}{dt} = -\Delta C_i \frac{1}{\tau} \qquad (3)$$

Comparing equation (3) to equation (1) gives

$$\frac{1}{\tau} = k_r + k_f\,(C_{Ni(NCS)^+,e} + C_{NCS^-,e})$$

with $C_{Ni(NCS)^+,e} = C_{NCS^-,e}$. As for the third component,

$$C_{Ni(NCS)_2,e} = C_{Ni(NCS)_2}\left(1 - \frac{\alpha}{100}\right)$$

where $C_{Ni(NCS)_2}$ is the total (associated plus dissociated) concentration and α is the percent dissociation. Assuming that K will be large, and therefore that α will be small,

$$K = \frac{k_f}{k_r} = \frac{C_{Ni(NCS)_2,e}}{[C_{Ni(NCS)^+,e}]^2}$$

$$= \frac{C_{Ni(NCS)_2}}{[C_{Ni(NCS)^+}\;]^2}$$

and the formula for the relaxation time becomes

$$\frac{1}{\tau} = k_r + 2(k_f k_r)^{1/2}\, C_{Ni(NCS)_2}^{1/2}$$

A plot of $1/\tau$ against , see the Fig. is linear with a slope of 4600 $M^{1/2}\ S^{-1}$ and an intercept of $k_r = 78\ S^{-1}$. The value of k_f is

$$k_f = \left[\frac{(slope)}{2k_r^{1/2}}\right]^2 = \left[\frac{4600}{(2)\,(78)^{1/2}}\right]^2 6.8 \times 10^4\ M^{-1}\ s^{-1}$$

and the value of K is

$$K = \frac{6.8 \times 10^4}{78} = 870.$$

Solution 24: When a reaction is displaced slightly from equilibrium, the return to equilibrium is first order in the displacement from equilibrium. The rate of approach to equilibrium is proportional to the displacement from equilibrium, and the relaxation time τ is used to characterize the rate of return to equilibrium. The forward rate $k_1\,[H_2O]$. The backward rate is

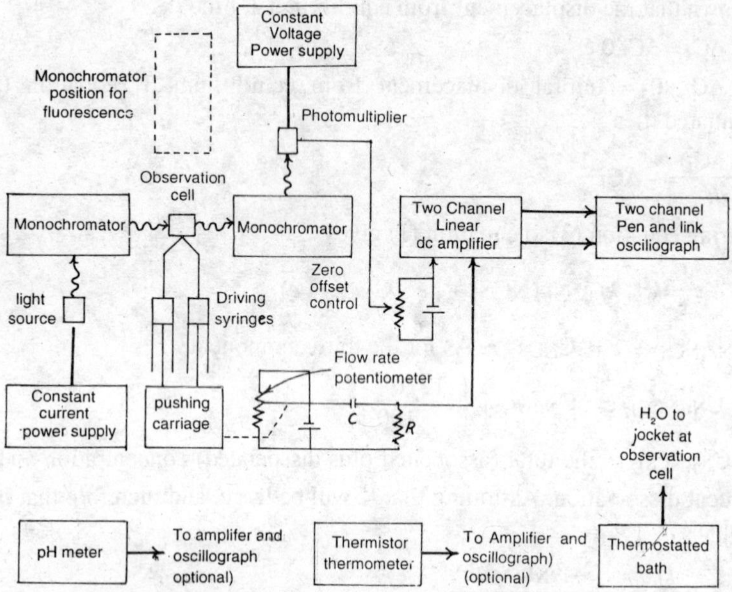

Block diagram of stopped-flow apparatus of type used to study rapid reactions in solution.

$$\frac{d\,(H_2O)}{dt} = k_2\,[H^+)\,(OH^-) - k_1\,(H_2O) \tag{1}$$

At equilibrium equation (1) changes to

$$0 = k_2\,(H^+)eq\,(OH\text{-})eq - k_1\,(H_2O)\,eq$$

In terms of the displacement, $\Delta\,(H_2O)$ from equilibrium, equation (1) can be written by introducing

$$(H^+) = (H^+)eq - \Delta\,(H_2O) \tag{2}$$

$$(OH^-) = (H^-)eq - \Delta\,(H_2O) \tag{3}$$

$$(H_2O) = (H_2O)eq - \Delta\,(H_2O) \tag{4}$$

Substituting equation (2), (3) and (4) into equation (1) gives

$$\frac{d\,(H_2O)}{dt} = k_2\,[(H^+)eq - \Delta\,(H_20)]\,[(OH^-)eq - \Delta\,(H_2O)]$$

$$-\,k_1\,[H_2O)eq + \Delta(H_2O)]$$

Now, make the assumption that $(H^+)eq\,(OH\text{-})eq$ displacement from equilibrium is small

$$\therefore \quad \frac{d[\Delta(H_2O)]}{dt} = k_2\,[(H+)_{eq}\,(OH-)_{eq} - (H+)_{eq}\,(+\Delta(H_2O))$$

$$- \Delta(H_2O)\,(OH^-)e2q + \Delta\,(H_2O)^2]$$

$$- k_1\,[(H_2O)eq + \Delta\,(H_2O)^2]$$

$$- k_1\,[(H_2O)eq + \Delta\,(H_2O)]$$

$$= - k_2\,[+(H^+)eq\,(+\,\Delta(H_2O)) + \Delta\,(H_2O)\,(OH^-)eq - \Delta\,(\,H_2O)^2] - k_1\Delta(H_2O)$$

$$+ [k_2\,(H^+)eq\,(OH-)eq\,(OH-)eq] - [k_1\,(H_2O)eq]$$

or $$\frac{d[\Delta(H_2O)]}{dt} = - \Delta(H_2O)\left\{k_2\,[(H^+)_{eq} + (OH^-)_{eq}] + k_1\right\} = \frac{-\Delta(H_2O)}{\tau}$$

where τ = the relaxation time. Observe that the term $[\Delta\,(H_2O)]^2$ has been neglected. Therefore

$$\frac{1}{\tau} = k_1 + k_2\,[(H^+)eq + (OH-)eq]$$

The equilibrium constant for the reaction is

$$K = \frac{(H^+)\,(OH^-)}{(H_2O)}$$

$$= \frac{k_1}{k_2}$$

$$= \frac{Kw}{(H_2O)}$$

where Kw = the dissociation constant for water.

Thus,

$$k = \frac{1.0 \times 10^{-14}}{55.5}$$

$$= 1.8 \times 10^{-16}$$

At equilibrium

$$k_1\,(H_2O)eq = k_2\,(H^+)eq\,(OH^-)eq$$

or $k_1 = k_2K(6)$

From equation (5)

$$\frac{1}{\tau} = k_2 + k_2\,[(H^+)eq + (OH^-)eq]$$

Therefore, since $(H^+)eq\,(OH^-)eq$ = Kw and $(H^+)eq = (OH^-)eq$

$$\frac{1}{\tau} = k_2 [\,K + \sqrt{Kw} + \sqrt{Kw}]$$
$$= k_2 [1.8 \times 10^{-16} + (2 \times 10^{-7})]$$
$$= 2 \times 10^{-7} k_2$$

and

$$k_2 = \frac{1}{(37 \times 10^{-6} (2 \times 10^{-7})]}$$
$$= 1.4 \times 10^{11} M^{-1} S^{-1}$$

Substitute k_2 into equation (6) to obtain

$$k_1 = (1.4 \times 10^{11})(1.8 \times 10^{-16})$$
$$= 2.5 \times 10^{-5} s^{-1}$$

SOLUTION 25:

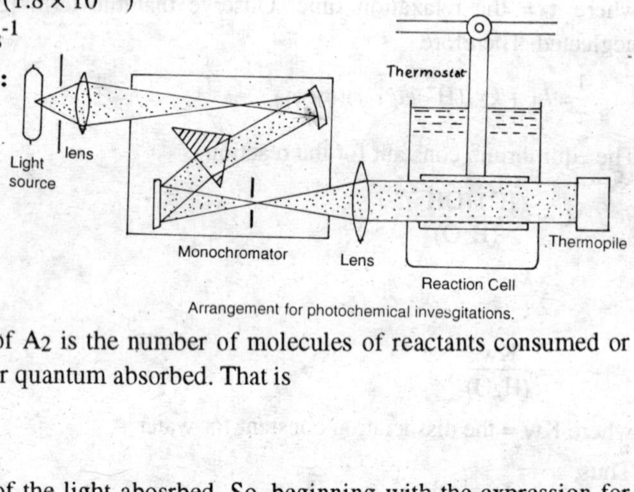

Arrangement for photochemical invesgitations.

The quantum yield of A_2 is the number of molecules of reactants consumed or product generated per quantum absorbed. That is

$$\frac{d\,(C_{A_2})/dt}{I}$$

where I = intensity of the light abosrbed. So, beginning with the expression for the rate of change of the concentration of A* gives

$$\frac{dC_{A*}}{dt} = k_1 I - k_2 C_{A*} C_A - k_3 C_{A*} \qquad (1)$$

where C = concentration

k_1, k_2 and k_3 = rate constants.

Assuming steady-state conditions, equation (i) is set equal to zero because the steady-state approximation assumes that the concentrations of certain intermediate substances reach constant values. That is

$$\frac{dC_{A*}}{dt} = K_1 I - K_2 C_A * C_{A*} \; K_3 C_{A*} = 0 \qquad (2)$$

Rearranging equation (2) to solve for C_{A*} yields

$$C_{A*} = \frac{k_1 I}{k_2 C_A + k_3} \quad (3)$$

The expression for the formation of A_2 is given by

$$\frac{d}{dt}(C_{A_2}) = k_2 C_{A*} C_A \quad (4)$$

The C_{A*} term in equation (4) is given by equation (3). Substituting this expression into equation (4) gives

$$\frac{d(C_{A_2})}{dt} = k_2 \frac{k_1 I}{k_2 C_A + k_3} C_A \quad (5)$$

Recall that the quantum yield is

$$\left(\frac{dC_{A_2}}{dt}\right)/I.$$

Hence, equation (5) changes to

$$\left(\frac{dC_{A_2}}{dt}\right)/I = \frac{k_2 k_1 C_A}{k_2 C_A + k_3}$$

Solution 26: A catalyst provides an alternative path for many reactions. When the reactions follow this alternative path, either the value of enthlpy, H or entropy S, or both are altered in such a way that it lowers the free energy barrier for the reaction. The reaction would then proceed at a more rapid rate.

the rates of the two reactions with subscript c to denote the catalyzed reactions are

$$r = r_O\, e^{G_c^{\circ} PT} \quad (1)$$

and $\quad r_c = r_o\, e^{G_c^{\circ} PT} \quad (2)$

where ro is a constant

Deviding equation (2) by equation (1) gives

$$\frac{r_c}{r} = \frac{e^{-G_c^{\circ}/RT}}{e^{-G^{\circ}/RT}} = e^{-G_C^{\circ}/RT} \quad e^{G^{\circ}/RT}$$

Taking the *ln* of both sides yields

$$ln\frac{r_c}{r} = \frac{G^{\circ} - G_C^{\circ}}{RT}$$

or

$$RT\, ln\frac{r_c}{r} = G^{\circ} - G_C^{\circ}$$

But $\quad G^{\circ} = H - TS^{\circ}$

Therefore, $Rt\ ln \frac{r_c}{r} = H - H_C - T(S^\circ - S_C^\circ)$ (3)

Assuming that S° is not affected by the catalyst, S° will be equal to S_c° and S° - $S_c^\circ = 0$. Therefore, equation (3) becomes

$$H - H_C = RT\ ln \frac{r_c}{r}$$

$$= \left(1.987 \frac{cal}{mole\ ^\circ K}\right)(533^\circ K)\ ln \frac{4.6}{0.015}$$

$$= 6063 \text{ cal/mole}$$

$$= 6.063 \text{ kcal/mole}$$

Solution 27: The reaction rate theory as put forward by Arrhenius, gives the variation of the rate constant, k with temperature as represented by the expression

$$\frac{d \ln k}{dT} = \frac{H}{RT^2} \qquad (1)$$

where H = the enthalpy increase per mole needed to raise the reactants to a state of activation. Intergrating equation (1) gives

$$\int_{k_1}^{k_2} ln\ k = \frac{H}{R} \int_{T_1}^{T_2} \frac{dT}{T^2}$$

Therefore

$$ln \frac{k_2}{k_1} = -H\left(\frac{1}{T_2} - \frac{1}{T_1}\right)$$

or

$$R\ ln \frac{k_2}{k_1} = -H\left(\frac{1}{T_2} - \frac{1}{T_1}\right) \qquad (2)$$

For the uncatalyzed reaction, equation (2) becomes

$$1.987 \times 10^{-3} \frac{kcal}{mole\ ^\circ K}\ ln \frac{0.804}{1.76 \times 10^{-2}} = -H\left(\frac{1}{573} - \frac{1}{473}\right)$$

$$1.987 \times 10^{-3} \frac{kcal}{mole\ ^\circ K}(3.82) = -H(-3.689 \times 10^{-4}\ ^\circ K^{-1})$$

Hence

$$H = \frac{1.987 \times 10^{-3} \frac{kcal}{mole\ ^\circ K}(3.82)}{3.689 \times 10^{-4}\ ^\circ K^{-1}}$$

$$= 20.6 \text{ kcal/mole}$$

For the catalyzed reaction, equation (2) gives

$$R\ ln \frac{(k_2)c}{(k_1)c} = -Hc\left(\frac{1}{T_2} - \frac{1}{T_1}\right)$$

The subscript c denotes catalyst.

Thus $1.987 \times 10^{-3} \frac{\text{kcal}}{\text{mole °K}}\ ln \frac{88.1}{6.10} = -H\ (3.689 \times 10^{-4})$

$$5.30 \times 10^{-3} \frac{\text{kcal}}{\text{mole °K}} = 3.689 \times 10^{-4}\ H.$$

and

$$H = \frac{5.30 \times 10^{-3} \frac{\text{kcal}}{\text{mole °K}}}{3.689 \times 10^{-4}\ °K}$$

$$= 14.4 \text{ kcal/mole}$$

The rate constant is given by

$$H = e^{-(H-TS°)/RT}$$

Taking the *ln* of both sides given

$$ln\ k = ln\ e^{-(H-TS°)/RT} = \frac{-H + TS}{RT}$$

or $RT\ ln\ k = -H + TS°$

Therefore

$$S° = R\ ln\ k + \frac{H}{T} \qquad (3)$$

For the uncatalysed reaction equation (3) becomes

$$S° = 1.987\ ln\ 1.76 \times 10^{-2} + \frac{20600 \text{ cal/mole}}{473\ °K}$$

$$= -8.027 + 43.551$$

$$= 35.5 \text{ eu}$$

For the catalysed reaction equation (3) gives

$$S° = 1.987\ ln\ 6.10 + \frac{14400 \text{ cal/mole}}{473\ °K}$$

$$= 34.0\ eu$$

Solution 28: $k = \frac{2.303}{t} \log\left(\frac{a}{a-x}\right) = \frac{2.303}{10} \log \frac{100}{24.8} = 0.2303 \log 4.03$

$$= 0.2303 \times 0.606 = 0/140 \text{ min}^{-1}$$

Solution 29 : Try second order.

$$k = \frac{2.303}{t(a-b)} \log \frac{b(a-x)}{a(b-x)}$$

$a = [NaClO]_0 = 0.003230$

$b = [kBr]_0 = 0.002508$

$a - b = 0.000722$

$$k = \frac{2.303}{3.65\,(7.22 \times 10^{-4})} \log \frac{2.508 \times 10^{-3}\,(0.003230 - 0.000560)}{3.230 \times 10^{-3}\,(0.002508 - 0.000560)}$$

$= 23.12$ litres mol^{-3} min^{-1}

$k = 23.42, 23.30, 23.52, 23.90, 23.80, 23.80$ average 23.62 liters $mole^{-1}$ min^{-1}

Second order (as shown by the approximate constancy of k)

Solution 30: Try first order.

$$k = \frac{1}{t} ln \left(\frac{c_o}{c}\right)$$

$$k = \frac{2.303}{120} \log \frac{11.45}{9.63} = 0.00144$$

$k = 0.00144, 0.00144, 0.00145, 0.00145, 0.00150, 0.00140,$ average 0.00145 min^{-1}

Second 31: For a second-order reaction

$$k = \frac{2.303}{t\,(a-b)} \log \frac{b\,(a-x)}{a\,(b-x)}$$

where

$a =$ concentration $S_2O_3^{2-}$ at time $t = 0$

$$= \frac{\text{meq iodine consumed at time } t}{V_{S_2O_3^{-2}}}$$

$$= \frac{37.63 \times 0.02572}{10.02} = 0.0967\,M$$

(At time $t = 0$ for this experiment, some $S_2O_3^{2-}$ had already been consumed, so its concentration was less than the initial 0.100 M.)

$b =$ concentration of n-PrBr at time $t = 0$

$$= \frac{\text{total meq thiosulfate consumed by } n\text{–PrBr}}{V_{n-PrBr}}$$

$$= \frac{(37.63 - 22.24) \times 0.02572}{10.02} = 0.0395\ M$$

We assume that at t = 78,840 sec equilibrium has been attained.

$(b - x)$ = concentration n-PrBr at time t

$(a - x)$ = concentration $S_2O_3^{2-}$ at time t

At t = 1110 sec

$$(b - x) = \frac{\text{meq } S_2O_3^{2-} \text{ consumed } at\ t = 1110 \text{ sec}}{V_{n-\text{PrBr}}}$$

$$= \frac{(35.20 - 22.24) \times 0.02572}{10.02} = \frac{12.96 \times 0.02572}{10.02}$$

$$(a - x) = \frac{\text{meq iodine consumed at } t = 1110 \text{ sec}}{V_{S_2O_3^{-2}}}$$

$$= \frac{35.20 \times 0.02572}{10.02}$$

$$\frac{b\,(a - x)}{a\,(b - x)} = \frac{0.0395\,(35.20 \times 0.02572)/(10.02)}{0.0967\,(12.96 \times 0.02572)/(10.02)}$$

$$= \frac{0.0395\,(35.20)}{0.0967\,(12.96)} = 1.110$$

$$a - b = 0.0967 - 0.0395 = 0.0572$$

$$k = 0.0363 \times \log 1.110 = 0.0363 \times 0.0453 = 0.001644$$

$$k = 0.001644, 0.001644, 0.001649, 0.001636, 0.001618, 0.001618$$

Average :

k = 0.001635 liter mole^{-1} sec^{-1}

Solution 32: For a second-order reaction in which the two reactants are present in equal concentrations

$$k = \frac{x}{ta\,(a - x)}$$

$$k = \frac{0.00876}{1200 \times 0.0198\,(0.01104)} = 0.0334$$

k = 0.0334, 0.0327, 0.0329, 0.0332, 0.0328, 0.0325: average 0.329 litre mole^{-1} min^{-1}

Solution 33 : For an nth-order reaction,

$$-\frac{dc}{dt} = kc^n$$

$$\log\left(-\frac{dc}{dt}\right) = \log k + n \log c$$

Thus

$$\frac{\log(-dc/dt)_2 - \log(-dc/dt)_1}{\log c_1 - \log c_2} = n$$

The rate of pressure increase is proportional to - *dc*/*dt*.

$$n = \frac{\log 7.4 - \log 0.73}{\log 2.153 - \log 0.433} = 1.44 \approx \frac{3}{2}$$

Solution 34: By inspection it can be seen that $t_{1/2}$ is directly proportional to *p*. Every time *p* is halved, $t_{1/2}$ is also halved. This suggests a zero-order reaction. For a zero-order reaction $-dx/dt = k$ or, in this case, $-dp/dt = k$, and on integration $k = P_i/2t_{1/2}$, where $P_{i.e.}$ is the initial pressure. k = 17.4, 17.6, 17.0; average 17.3 torr min^{-1}

Solution 35: *A* is directly proportional to concentration.

Plot $\log(A - A_\infty)$ vs *t*. Graph is a straight line with slope = -0.0563; $k = -2.303 \times$ slope $= 0.130\ \text{min}^{-1}$.

Solution 36: x = fraction converted $= \dfrac{\alpha_0 - \alpha_t}{\alpha_0 - \alpha_\infty} = \dfrac{11.20 - 10.35}{11.20 - (-3.37)}$

$$= 0.0583 \text{ at } t = 1035 \text{ sec.}$$

Try first order.

$$k = \frac{2.303}{t}\log\frac{1}{1-x} = 5.80 \times 10^{-5}$$

$$k_{av} = 5.80 \times 10^{-5}\ \text{sec}^{-1}$$

First order.

Solution 37: Try first order

$$\alpha = 18.48 - 12.29 = 6.19$$

$$a - x = 18.05 - 12.29 = 5.76$$

$$k = \frac{2.303}{t}\log\left(\frac{a}{a-x}\right) = \frac{2.303}{30}\log\frac{6.19}{5.76} = 2.40 \times 10^{-3}\ \text{min}^{-1}$$

k = 2.40, 2.46, 2.49, 2.49; average $2.46 \times 10^{-3}\ \text{min}^{-1}$.

Solution 38: (a) $K = \dfrac{[NOH_2]}{[NO]\,[H_2]}$

$$\frac{d[NOH]}{dt} = 2k[NOH_2]\,[NO] = 2kK[NO]^2[H_2]$$

$$\text{(b) } K' = \frac{[N_2O_2]}{[NO]^2}$$

$$\frac{d[NOH]}{dt} = 2k[N_2H_2]\,[H_2] = 2k'K'[NO]^2[H_2]$$

Solution 39: (a) $t_{1/2} = \frac{1}{k} ln \left(\frac{a}{a/2}\right) = \frac{1}{k} ln\, 2 = \frac{0.693}{k}$

$$\text{(b) } t_{1/2} = \frac{1}{k}\left(\frac{1}{a/2} - \frac{1}{a}\right) = \frac{1}{ka}$$

$$\text{(c) } t_{1/2} = \frac{1}{2k}\left(\frac{1}{(a/2)^2} - \frac{1}{a^2}\right) = \frac{3}{2ka^2}$$

Solution 40: Plot log k vs $\sqrt{I}$ and exprapolate to $\sqrt{I} = 0$.

5 + log k	$\sqrt{I}$
1.7640	1.53
1.7130	2.37
1.6800	2.85
1.6467	3.35
1.6418	3.42
1.5990	4.11

Exptrapolated value:

$5 + \log k_0 = 1.858$, $k_0 = 7.21 \times 10^{-4}$

Solution 41: $2 = \frac{k_{305}}{k_{295}} = \frac{e^{-E/305\,R}}{e^{-E/295\,R}}$

$$ln\, 2 = -\frac{E}{R}\left(\frac{1}{305} - \frac{1}{295}\right) = -\frac{E}{R}\left(\frac{-10}{305 \times 295}\right)$$

$$0.693 = \frac{10E}{1.987 \times 10^{-3} \times 305 \times 295}$$

$$E = 12.4 \text{ kcal mole}^{-1}$$

Solution 42: $k = Ae^{-E/RT}$

$$E = \frac{RT_1\,T_2 \ln (k_2/k_1)}{T_2 - T_1}$$

$$\frac{2.303\, RT_1\, T_2}{T_2 - T_1} = \frac{4.576 \times 910 \times 586}{910 - 586} = 7530 \text{ cal mole}^{-1}$$

$$E_{max} = 7530 \log \left(\frac{1.85 \times 10^{11}}{1.58 \times 10^{10}}\right)$$

$$= 8046 \text{ cal mole}^{-1} \text{ of activated complex}$$

$$E_{min} = 7530 \log\left(\frac{1.69 \times 10^{11}}{1.68 \times 10^{10}}\right)$$

$$= 7549 \text{ cal mole}^{-1}$$

(Wilson, using data at temperatures down to 300°K, obtains $E = 7900 \pm 200$ cal mole^{-1}).

Solution 43 : It can be seen from inspection of the table that doubling the MnO_4^- concentration doubles the reaction rate, while decreasing the Cr^{3+} concentration by one half decreases the reaction rate by one half. Therefore the reaction is first order with respect to Cr^{3+} and first order with respect to $Mn4^-$.

Solution 44:

	Percent complete
t	*(100 $V_{HCl,t}/V_{HCl,\infty}$)*
30	55.9
44	68.3
62	80.8
83	87.0
111	93.2
∞	100.0

$$k_1 = \frac{1}{\tau} \times 2.303 \log\left(\frac{a}{a-x}\right) = \frac{2.303}{30} \log \frac{100}{44.1} = 0.0273 \text{ sec}^{-1}$$

$k_1 = 0.0273, 0.0262, 0.0266, 0.0249, 0.0242$; average 0.0258 sec^{-1}

The approximate constancy of k_1 shows that the reaction is first order.

Solution 45 : EtOAc + NaOH → EtOH + NaOAc

Try second orer.

$$k = \frac{2.303}{t(a-b)} \log\left[\frac{b(a-x)}{a(b-x)}\right]$$

$a = 0.0200 = [\text{NaOH}]_0$

$b = 0.0100 = [\text{EtOAc}]_0$

At $t = 15$ min

$[\text{NaOH}] = (50.0 - 7.35) \times 0.0200 \times 10^{-3}$ mole/0.05000 liter

$= 0.01706 \text{ N} = a - x$

$x - a - 0.01706 = 0.0200 - 0.01706 = 0.00294$

(EtOAc] $b - x = 0.0100 - 0.00294 = 0.00706$

$$k = \frac{2.303}{15(0.0100)} \log \left[\frac{0.0100(0.01706)}{0.0200(0.00706)}\right] = 1.260 \text{ liter mole}^{-1} \text{ min}^{-1}$$

$k = 1.260, 1.136, 1.087, 1.053, 1.039$, average 1.097 liters mole^{-1} min^{-1}

Second order (as shown by the approximate constancy of k)

Solution 46: The data do not fit first, second, or third order. By polotting ml $FeSO_4$ consumed, which is proportional to meq $K_2S_2O_8$ remaining, it can be seen that the concentration of $K_2S_2O_8$ varies linearly with time. Thus the reaction rate is independent of $K_2S_2O_8$ concentration and the reaction is zero order. Alternatively we could calculate meq $K_2S_2O_8$ concentration and can vary continuously from zero to one. In the concentration range referred to here the order is zero.

Solution 47: The [$CHCl_3$] and [NaOMe] data are not needed to solve the problem.

Method 1. Plot log ($\nu_\infty - \nu$) vs t. If first order, a straight line will be obtained. The value of k can be calculated from the slope of this line.

Method 2. Sample calculation for $t = 4$ min:

$$k = \frac{2.303}{t} \log \frac{\vartheta_\infty - \vartheta_0}{\vartheta_\infty - \vartheta}$$

$$= \frac{2.303}{4} \log \frac{14.27}{12.95} = 0.0243 \text{ min}^{-1}$$

$k = 0.0243, 0.0241, 0.0236, 0.231, 0.0232, 0.0233, 0.0241$, average 0.0237 min^{-1}

The approximate constancy of k shows that the reaction is first order.

Solution 48:

t (min)	Percent complete (100 $\nu FeSO_4.t/\nu FeSO_4.\infty$)
0.31	42.4
0.51	54.6
0.65	60.6
0.86	66.7
1.01	69.7

1.15	72.7
1.36	75.8
1.58	78.8
2.19	83.6
∞	100

$$k = \frac{1}{at}\left(\frac{x}{a-x}\right)$$

where a = 100 per cent, x = percent complete.

$10^2 k$ = 2.37, 2.36, 2.36, 2.33, 2.32, 2.30, 2.35, 2.33; average $k = 2.33 \times 10^{-2}$ liter $mole^{-1}$ min^{-1}

Solution 49: (a) $2AsH_3\,(g) \rightarrow 2As(s) + 3H_2(g)$

$P_0 - 2x$ $\qquad$ $3x$

Let

P_0 = initial pressure (at time t_0) 39.2 cm

$P_0 - 2x$ = partial pressure of AsH_3 at time t

Then $3x$ = partial pressure of H_2 at time t. Therefore

$$P_{total} = P_0 - 2x + 3x = P_0 + x$$

at time t. At t = 4.33 hr,

$$P_{total} = 39.2 + x = 40.3,\ x - 1.1$$

Therefore,

$$P_{ASH^3} = P_0 - 2x = 39.2 - 2(1.1) = 37.0 \text{ cm}$$

t	0	4.33	16	25.5	37.66	44.75
log P_{ASH3}	1.593	1.568	1.48	1.430	1.332	1.299

If the reaction is first order with respect to AsH_3, then

$$\frac{dP_{ASH^2}}{dt} = -kP_{ASH3}$$

Thus

$$\frac{d\,P_{ASH_2}}{P_{ASH_2}} = -k\,dt$$

and, on integration as an indefinite integral,

$$ln\,P_{SH_3} = -\,kt + \text{constant}$$

$$2.303 \log P_{ASH_3} = -kt + \text{constant}'$$

$$\log P_{ASH_3} = -\frac{ko}{2.303} + \text{constant}'$$

This is the equation of a straight line with slope = - $k/2.303$. Plot log vs t. A straight line is obtained. Therefore the reaction is first order.

(b) $$\text{Slope} = \frac{-k}{2.303} = 6.58 \times 10^{-3}$$

$$k = 1.52 \times 10^{-2}\ \text{hr}^{-1}$$

Solution 50: At 5 hr, the fraction of methyl tosylate unchanged = (0.746 - 0.420) / (0.791 - 0.420) = 0.879.

$$k = \times = 7.17 \times 10^{-3}\ \text{sec}^{-1}$$

$$k_{avg} = 7.77 \times 10^{-6}\ \text{sec}^{-1}$$

Solution 51: Let A be the reacting gas, B the inert gas, and C and D the products.

$$A + A \underset{k_{-2}}{\overset{k_1}{=\!=}} A^* + A$$

$$A + B \underset{k_{-2}}{\overset{k_1}{=\!=}} A^* + B$$

$$A^* \overset{k_3}{=\!=} C + D$$

$$\frac{d[A^*]}{dt} = k_1\,[A]^2 - k_1[A^*][A] + k_2\,[A][B] + k_2\,[A][B] - k_{-2}\,[A^*][B] - k_3[A^*]$$

$$\approx 0$$

$$\frac{d[C^*]}{dt} = k_3\,[A^*] = \frac{k_3(k_1[A]^2 + k_2A][B]}{k_{-1}[A] + k_{-2}[B] + k_3}$$

Solution 52 : (a) $\frac{d[O]}{dt} = k_1\,[P_3] - k_{-1}\,[O][O_2] - k_2\,[O][O_2] - k_2[O][O_3] \approx 0$

$$[O] = \frac{k_1O_3]}{k_{-1}[O_2] + k_2[O_2]}$$

$$\frac{d[O_3]}{dt} = -k_1\,[O_3] + k_{-1}\,[O]\,[O_2] - k_2\,[O][O_3]$$

$$= -k_1\,[O_3] + \frac{k_{-1}k_1[O_3][O_2] - k_1k_2[O_3]^2}{k_{-1}[O_2 + k_2[O_3]}$$

$$= \frac{-2k_1k_2[O]^2}{k_{-1}[O_2] + k_2[O_3]}$$

(b) If

$$k_{-1}[O_2] >> k_2[O_3]$$

$$\frac{d[O_2]}{dt} \approx \frac{-2k_1k_2[O_3]^2}{k_{-1}[O_2]}$$

Solution 53 : (a) Mechanism (i):

$$\frac{d[HI]}{dt} = 2k[H_2][I_2]$$

Mechanism (ii):

$$\frac{d[I]}{dt} = 2k_1[I_2] - 2k_2[I]^2 - 2k_3[I^2][H_2] \approx 0$$

$$[I^2] = \frac{k_1[I_2]}{k_2 + k_2[H_2]}$$

$$\frac{d[HI]}{dt} = 2k_3[I]^2[H_2] = \frac{2k_1k_2[H_2][I_2]}{k_2 + k_3[H]}$$

(b)

$$\frac{[I]^2}{[I_2]} = K[I^2]\,K[I_2]$$

$$\frac{d[HI]}{dt} = 2k_3K[H_2][I_2]$$

This result agrees with that of (a) if $k_3[H_2]\ k_2$, that is, if reaction (3) is much slower than reaction (2).

(c)

$$ln\,k_3 = ln\,A - \frac{E_a}{RT}$$

$$ln\left(\frac{k_3,T_2}{k_3,T_2}\right) = \frac{E_a}{R}\left(\frac{1}{T_1} - \frac{1}{T_2}\right) = \frac{E_a}{R}\left(\frac{T_2 < T_1}{T_1T_2}\right)$$

$$Ea = \frac{RT_1T_2}{T_2 - T_1}\,ln\left(\frac{k_3,T_2}{k_3,T_1}\right)$$

$$= \frac{2.303 \times 1.987 \times 10^{-3} \times 417.9 \times 520.1}{520.1 - 417.9}$$

$$\times \log\left(\frac{4.0 \times 10^5}{1.12 \times 10^5}\right)$$

$= 5.38$ kcal mole^{-1} of activated complex

Solution 54: (a) $\frac{d[NO_3-]}{dt} = k_1[NO_{2-}][O_2] + k_2[O][NO_{2-}]$ (1)

$$\frac{d[O]}{dt} = k_1 \, [NO_{2^-}] \, [O_2] - k_2 \, [O] \, [NO_{2^-}] - k_3 \, [O^2]$$

Assuming steady state of [O], then from equation (2),

$$[O] = \frac{k_1[NO_2-][O_2]}{k_3[O] + k_2[NO_2-]}$$

Substituting (3) into (1)

$$\frac{d[NO_3-]}{dt} = k_1 \, [NO_{2^-}] \, [O_2] \left(1 + \frac{k_2[NO_2-]}{k_3[O] + k_2[NO_2-]}\right)$$

(b) If reaction (3) is much slower than reaction (1), then k_3 [O] <<$k_2[NO_{2^-}]$, and

$$\frac{d[NO_3-]}{dt} = k_1 \, [NO_{2^-}][O_2] \left(1 + \frac{k_2[NO_2-]}{k_2[NO_2-]}\right)$$

$$= 2k_1 \, [NO_{2^-}] \, [O_2]$$

Solution 55 : (a) $= K \dfrac{[BH-][O_2 = NH-]}{[O_2NNH_2][B]}$

$$\frac{d[O_2NNH_2]}{dt} \; -k_2 \, [O_2NNH-] \, \frac{k_2K[O_2NNH_2][B]}{[BH-]}$$

(b) $\dfrac{d\,[O_2NNH_2]}{dt} \; -k_1 \, [O_2 \, NNH_2] \, [B]$

Both assumptions are consistent with the first-order rate law. They can be distinguished by performing experiments in which [BH-] is varied while [O_2 NH_2] and [B] are held constant. If (a) is correct, the rate should be inversely proportional to [BH^+]. If (b) is correct, the rate should be constant in these experiments. (It turns out that (b) is correct.)

Solution 56: (a) $\dfrac{d\,[NO_2-]}{dt} = k_1 \, \phi \, [NO_{3^-}] - k_2 \, [O][NO^2-] + k^3 \, [O] \, NO_{\overline{3}}]$

Since ϕ is the incident (rather than the absorbed) dose rate, the rate of reaction (1) is given by $k_1\phi$ $[NO_{3^-}]$ (not $k_1\phi$), provided that only a small fraction of the incident radiation is absorbed. Maing the steady state assumption about O, we obtain

$$\frac{d[O]}{dt} = k_1 \, [NO_{3^-}] \, \phi - k_2 \, [O][NO_{2^-}] - k_3 \, [O] \, [NO_{3^-}] = 0$$

$$[O] = \frac{k_1[NO_2-] \, [NO_3-]\varphi}{k_2[NO_2-] + k_3[NO_3-]}$$

and therefore

$$\frac{d\,[NO_2-]}{dt} = k_1\phi \, [NO_3-] + \frac{-k_1k_2[NO_2-] \, [NO_3-]\varphi + k_1k_2[NO_3-]^3\varphi}{k_2[NO_2-] + k_3[NO_3-]}$$

$$= \frac{-k_1k_2[NO_2-][NO_3-] + k_1k_3\,\varphi\,[NO_3-]^2 \quad -k_1k_2[NO_2-][NO_3-] + k_1k_3\,\varphi\,[NO_3-]^2\,\varphi}{k_2[NO_2-] + k_3[NO_3-]}$$

$$= \frac{2k_1k_2\,\varphi\,[NO_3-]}{k_2[NO_2-] + k_3[NO_3-]}$$

(b) Rearranging the last expression we obtain

$k_2\,[NO_{2-}]\,d\,[NO_{2-}] + k_3\,[NO_{3-}]\,d\,[NO_{2-}] = 2k_1k_3\phi\,[NO_{3-}]^2\,dt,$

which, upon integration between limits 0 to $[NO_{2-}]$ for $[NO2-]$ and between 0 and t, gives

$$k_2\,[NO_{2-}]^2 + k_3\,[NO_{3-}]\,[NO2-] = 2k_1k_3\phi\,[NO_{3-}]^2\,t$$

Solution 57 : $\frac{d[CH_3]}{dt} = k_1\,[CH_4] - k_2\,[CH_3][CH_4] + k_3\,[H]\,[CH_4]$

$-k_4\,[H]\,[CH_3]\,[M] = 0$

$$\frac{d[H]}{dt} = k_1\,[CH_4] + k_2\,[CH_3]\,[CH_4] - k_3\,[H][CH_4]$$

$$-\,k_4\,[H]\,[CH_3]\,[M] = 0$$

Adding

$$k_1\,[CH_4] - k_4\,[H]\,[CH_3]\,[M] = 0$$

Subtracting

$$-k_2\,[CH_3]\,[CH_4] + k_3\,[H][CH_4] = 0$$

$$[H]\,[CH_3] = \frac{k_1[CH_4]}{k_4[M]}$$

$$\frac{[H]}{[CH_3]} = \frac{k_2}{k_3}$$

Dividing

$$[CH_3]^2 = \frac{k_1k_3[CH_4]}{k_2k_4[M]}$$

$$\frac{d[C_2H_6]}{dt} = k_2\,[CH_3]\,[CH_4] = \left(\frac{k_1k_2k_3[CH_4]^3}{k_4[M]}\right)^{1/2}$$

Solution 58 :

(a) $$-\frac{dc}{dt} = kc^{-1}$$

$$-cdc = k\,dt$$

$$\frac{1}{2}(c_0{}^2 - c^2) = kt$$

$$c = (c_0{}^2 - 2kt)^{1/2}$$

(b) $$c = 0.1\ c_0$$

$$c^2 = 0.01\ c_0{}^2$$

$$t = \frac{c_0^2 - c^2}{2k} = \frac{0.99 c_0^2}{2k}$$

(c) $c = 0$ when $t = c_0{}^2/2k$. The reaction reaches completion at this time. In a first-order reaction $-dc/dt = kc$. As c decreases the rate of decrease ($-\,dc/dt$) also decreases, so that the reaction becomes slower and slower without ever attaining completion. In this -1 order reaction, becoming inifite as c appproaches zero. The reaction accelerates as it approaches completion and comes to a sudden stop when the last trace of reactant is consumed.

Solution 59 : (a) $$\frac{d[CI]}{dt} = k_1\ [NO_2Cl] - k_2\ [NO_2Cl]\ [Cl] \approx 0$$

$$[Cl] = \frac{k_1}{k_2}$$

$$\frac{d[NO_2CI]}{dt} = -k_1\ [NO_2Cl] - k_2\ [NO_2Cl]\ [Cl] = -2k_1\ [NO_2Cl]$$

(b) According to the Lindemann theory of unimolecular reactions, the NO_2Cl molecule must be activated by collision before it can decompose. It can also be deactivated by collision. We treat the activated molecule NO_2Cl^* as a distinct species.

$$2NO_2Cl \underset{k_{3-}}{\overset{k_3}{\leftrightarrow}} NO_2Cl + NO_2Cl^*$$

$$NO_2Cl^* \overset{k_1}{\rightarrow} NO_2 + Cl$$

$$NO_2Cl + Cl \overset{k_2}{\rightarrow} NO_2 + Cl_2$$

The steady state approximation can now be applied to both Cl and NO_2Cl^*:

$$\frac{d[CI]}{dt} = k_1'\ [NO_2Cl^*] - k_2\ [NO_2Cl]\ [Cl] \approx 0$$

$$\frac{d[NO_2CI*]}{dt} = k_3\ [NO_2Cl^2] - k_{-3}\ [NO_2Cl]\ [NO_2Cl^*]$$

$$-k_1'\ [NO_2Cl^*] \approx 0$$

$$[NO_2Cl^*] = \frac{k_3[NO_2Cl]^2}{k_{-3}[NO_2Cl] + k_1}$$

$$[Cl] = \frac{k_1[NO_2Cl^*]}{k_2[NO_2Cl]} = \frac{k_1k_3[NO_2Cl]}{k_2(k_{-3}[NO_2Cl] + k_1)}$$

$$= -k_3\,[NO_2Cl^2] + k_3\,[NO_2Cl]\,[NO_2Cl^*]$$

$$-k_2\,[NO_2Cl^*] - k_2\,[NO_2Cl]\,[Cl]$$

$$= \frac{-2k_1'k_3[NO_2Cl]^2}{k_{-3}[NO_2Cl] + k_1}$$

At high concentration k_{-3} $[NO_2Cl]$ k_1' and

$$\frac{d[NO_2Ci]}{dt} = \frac{-k_i'k_3[NO_2Ci]}{k_{-3}}$$

This result agrees with (a), provided that $k_1 = k_1'\,k_{-3}$. At low concentration $k_{-3}[NO_2Cl]$ k_1' and

$$\frac{d[NO_2Ci]}{dt} = -2k_3\,[NO_2Cl]^2$$

In the latter case activation by collision is the rate- determining step.

Solution 60 : (a) $\frac{d[HNO_3]}{dt} = -ka\,[HNO3] + kb\,[HO]\,[NO2] - kc\,[HO]\,[HNO3]$

$$\frac{d[HN]}{dt} = ka\,[HNO3] - kb\,[HO]\{NO2] - kc\,[HO]\,[HNO3] = 0$$

$$[HO] = \frac{k_a[HNO_3]}{k_b[NO_2] + k_c[HNOsusb3]}$$

$$\frac{d[HNO_3]}{dt} = -k_a\,[HNO_3] + \frac{(k_v[NO_2] - k_c[HNO_3]\,k_a[HNO_3]}{k_v[NO_2] + k_c[HNO_3]}$$

$$= \frac{-k_ak_b[HNO_3]\,k_v\,[NOsub2] - k_akcubc\,[HNO_3]^2 + k_ak_b[HNO_3]k_b\,[NO_2] - k_ak_c\,[HNO_3]^2}{k_b[NO_2] + k_c[NO_3]}$$

$$= \frac{-2k_akcubcHNO_3]^2}{k_v[NO_2] + k_c[HNO_3]}$$

$$= -2k_a\,[HNO_3]\,\frac{1}{1 + (k_b[NO_2]/(k_c[HNO_3]}$$

(b) If the consumption of NO_2 is very rapid, then $[NO_2] \approx 0$ therefore

$$\frac{k_b[NO_2]}{[k_c[HNO_3]} << 1$$

and

$$= \frac{d[HNO_3]}{dt} - 2k_a\,[HNO_3]\,.\,\frac{1}{1+0} = -2k_a\,[HNO_3]$$

k_d is not needed.

Solution 111 : $\frac{d[NpO_2^{\ +}]}{dt} = k_1\,[HO_{2^-}]\,[NpO_1^{2+}] + k_2\,[HO_2]\,[NpO_2^{+2}]$

$-k_3\,[HO_2][NpO_2^0]$

$\frac{d[HO_2]}{dt} = k_1\,[HO_{2^-}][NpO_2^{2+}] - k_2\,[HO_2[NpO_2^{2+}]$

$-k_3\,[HO_2]\,[NpO_2^+] = 0$

$$[HO_2] = \frac{k_1[HO_2^{\ -}]\,[NpO_2^{2+}]^2\,[HO_2^{\ -}]}{k_2[NpO_2^{2+}] + k_3[NpO_2^{\ +}]}$$

$$\frac{d[NpO_2^{\ +}]}{dt} = k_1\,[HO_{2^-}][NpO_2^{2+}] + \frac{k_1k_2\,[NpO_2^{2+}]^2\,[HO_2^{\ -}]}{k_2[NpO_2^{2+}] + k_3[NpO_2^{\ +}]}$$

$$-\frac{k_1k_3[HO_2^{\ -}][NpO_2^{2+}]}{k_2[NpO_2^{2+}] + k_3[NpO_2^{\ +}]}$$

$$= \frac{k_1[HO_2^{\ -}](k_3[Np_pO_2^{2+}]\,[NpO_2^{2+}] + k_2[NpO_2^{\ 2+}] - k_3[NpO_2^{\ +}][NpO_2^{2+}]}{k_2[NO_2-] + k_3[NO_3-]}$$

$$= \frac{k_1[HO_2^{\ -}]\,2k_2[NpO_2^{2+}]^2}{k_2[NpO_2^{2+}] + k_3[NpO_2^{\ +}]}$$

$$= \frac{2k_1[HO_2^{\ -}]\,[NpO_2^{2+}]}{1 + (k_3[NpO_2^{2+}])/(k_2[NpO_2^{2+}])}$$

$$K = \frac{[H^+]\,[HO_2^{\ -}]}{[H_2O_2]}$$

$$[HO_{2^-}] = \frac{K[H_2O_2]}{[H^+]}$$

$$\frac{d[NpO_2^{\ +}]}{dt} = \frac{2Kk_1[NpO_2^{2+}][H_2O_2]}{[H^+]\{1 + (k_3[NpO_2^{\ +}])/(k_2[NpO_2^{2+}])\}}$$

Solution 62 : (a) $\frac{d[NO_3]}{dt} = k_1\,[N_2O_5] - k_{-1}\,[NO_2]\,[NO_3] - k_2\,[NO_2]\,[NO_3] \approx$

0

$$[NO_3] = \frac{k_1[N_2O_5]}{(k_{-1}+k_2)[NO_2]}$$

$$\frac{d[NO]}{dt} = k_2\,[NO_2][NO_3] - k_3\,[NO]\,[N_2O_5] \approx 0$$

$$[NO] = \frac{k_2[NO_2][NO_3]}{k_3 N_2O_5]} = \frac{k_1 k_2}{k_3(k_{-1}+k_2)}$$

$$\frac{d[N_2O_5]}{dt} = -\,k_1\,[N_2O_5] + k_{-1}\,[NO_2][NO_3] - k_3\,[NO][N_2O_5]$$

$$= \left\{-k_1 + \frac{k_1(k_{-1}-k_2)}{k_{-1}+k_2}\right\}[N_2O_5]$$

(b) When NO_2 and NO_3 collide with the N of NO_2 meeting an O and NO_3, the product (if any) will be N_2O_5 (that is, O_2 $NONO_2$). When they collide with O's meeting, the products could be NO (from NO_2) + NO_2 (from NO_3) + O_2 (1 atom from each). Since there are half as many N atoms as 0 atomes in NO_2, the first kind of collision (needed for reaction (-1) should be approximately half as probable as the second kind (needed for reaction (2)). Then the frequency factors of reactions (-1) and (2) should be approximately in the ration 1:2:

$$\frac{k_{-1}}{k_2} = \frac{A_{-1}e^{-E_{-1}/RT}}{A_2 e^{-E_2/RT}} = \frac{e^{-0}}{2e^{-5/RT}}$$

$$= \frac{1}{2} \times 10^{5/(2.303} \times 1.987 \times 10^{-3} \times 320)$$

$$= 10^3$$

(c) $N_2O_5 = NO_2 + NO_3$

$$\frac{[NO_2][NO_3]}{[N_2O_5]} = K = \frac{k_1}{k_{-1}}$$

$$[NO] = \frac{k_2[NO_2][NO_3]}{k_3[N_2O_5]}$$

$$= [-k_1 + (k_{-1} - k_2)\,k]\,[N_2O_5]$$

This result is the same as that of (a) if $k_2 k_{-1}$, that is, if most NO_2 + NO_3 reactions result in N_2O_5. In this case reaction (2) can be considered as merely a lsow leakage out of the equilibrium system. The calculation in (b) supports this assumption.

Solution 63 : $\frac{d[Ci]}{dt} = 2k_1 [Cl_2] - 2k_2[Cl]^2 - k_3 [Cl][CO] + k_4 [COCl]$

$+ k_5 [COCl][Cl_2] = 0$ (1)

$$\frac{d[COCl]}{dt} = k_3 [C] [CO] - k_4 [COCl] - k_5 [COCl] [Cl_2] = 0 \quad (2)$$

Adding Equation (1) and (2)

$2k_1 [Cl_2] - 2k_2 [Cl]^2 = 0$

$$[Cl] = \left(\frac{k_1[Cl_2]}{k_2}\right)^{1/2}$$

From equation (2)

$$[COCl] = \frac{k_3[Cl][CO]}{k_4 + k_5[Cl_2]}\sqrt{x}$$

$$= \frac{k_3 k_1^{1/2}[Cl\sin 2]^{1/2}[CO]}{k_2^{1/2}(k_4 + k_5[Cl_2])}$$

Then

$$\frac{d[COCl_2]}{dt} = k_5 [COCl][Cl_2] = \frac{k_5 k_3 k_1^{1/2}[Cl_2]^{3/2}[CO]}{k_2^{1/2}(k_4 + k_5[Cl_2])} \quad (3)$$

(b) If k_4 [COCl] k_5 [COCl][Cl_2, equation (3) becomes

$$\frac{d[COCl_2]}{dt} = \frac{k_5 k_3 k_1^{1/2}[Cl_2]^{3/2}[CO]}{k_2^{1/2} k_4} \quad (4)$$

(c) $$\frac{[Cl]^2}{[Cl_2]} = K_1 [Cl] = K_1^{1/2} [Cl_2]^{1/2}$$

$$\frac{[COCl]}{[cl][CO]} = K_3 [COCl] = K_3 [Cl][CO]$$

$$= K_3 K_1^{1/2} [Cl_2]^{1/2} [CO]$$

$$= K_3 K_1^{1/2} [Cl_2]^{1/2} [CO]$$

$$\frac{d[COCl_2]}{dt} = k_5 [COCl][Cl_2] = k_5 K_3 K_1^{1/2} [Cl_2]^{3/2} [CO] \quad (5)$$

Equation (5) agrees with equation (4) if $K_1 = k_1/k_2$ and $K_3 = k_3/k_4$. This is the familiar relation between an equilibrium constant and the rate constants of the forward and equivalent to saying, as in (b), that they are much faster than reaction (5) and their equilibrium is not significantly perturbed by the relatively slow leadage of COCl and Cl_2 out of the equilirbium system.

Solution 64 : Let a = [A], b = [B], and c = [C]. The initital value of a (at t = 0) is a_0, whcih is assumed to be known. The rate equations are (in the notation $\dot{a}$ = da/dt, etc.)

$$\dot{a} = -k_1 a \quad (1)$$

$$\dot{b} = k_1 a - k_2 b \quad (2)$$

$$\dot{c} = k_2 b \quad (3)$$

For conservation $a + b + c = a_0$ or $a = a_0 - b - c$. Then

$$\dot{b} = k_1 (a_0 - b - c) - l_2 b \quad (4)$$

$$\ddot{b} = k_1 - k_2$$

$$\ddot{b} = k_1 (-\dot{b} - \dot{c}) - k_2 \dot{b}$$

Using (3)

$$\ddot{b} = k_1 (-\dot{b} - k_2 b) - k_2$$

or

$$\ddot{b} + (k_1 + k_2)\dot{b} + k_1 k_2 b = 0 \quad (5)$$

We now have one linear homogeneous differential equation in the variable b. The general solution of this equation is

$$b = \alpha e^{+m_1 t} + \beta e^{+m_2 t} \quad (6)$$

where α and β are constants of integration, and μ_1, μ_2 are the two distinct roots of the equation

$$m^2 + (k_1 + k_2)m + k_1 k_2 = 0$$

$$m_1 = -k_1, m_2 = -k_2$$

If $k_1 = k_2$, the roots are not distinct and (6) is not the most general solution. We first consider

Case I. $k_1 \neq k_2$

$$b = \alpha e^{+k_1 t} + \beta e^{+k_2 t}$$

When $t = 0$, $b = 0$; $0 = \alpha + \beta$, $\beta = -\alpha$

$$b = \alpha (e^{-k_1 t} - e^{-l_2 t}) \quad (7)$$

To find α, we start with (4)

$$\dot{b} = k_1 (a_0 - b - c) - k_2 b$$

When $t = 0$, $b = c = 0$. Then $= k_1 a_0$. From (7)

$$\dot{b} = \alpha(-k_1 e^{-k_1 t} + k_2 e^{-k_2 t})$$

when $t = 0$, $= \alpha (k_2 - k_1)$

$$\alpha (k_2 - k_1) = k_1 a_0$$

$$\alpha = \frac{k_1 a_0}{k_2 - k_1}$$

When b is a maximum, $= 0$ and b 0.

$$\dot{b} = \alpha (-k_1 e^{-k_1 t} + k_2 e^{-k_2 t}) = 0$$

$$-k_1 e^{-k_1 t} + k_2 e^{-k_2 t}$$

$ln\ k_1 - k_1 t = ln\ k2 - k_2 t$

$$\max = \frac{\ln(k_1/k_2)}{k_1 - k_2}$$

(Can this expression ever be negative ?)

$$\ddot{b} = -(k_1 + k_2)\dot{b} - k_1 k_2 b \text{ (from 5))}$$

$$= -k_1 k_2 b < 0$$

when $\dot{b} = 0$.

Therefore b has a maximum when $t = t_{\mathbf{max}}$, not a minimum or a horizontal point of inflection.

$$b_{\mathbf{max}} = \alpha \left\{ \exp\left[\frac{-k_1 \ln(k_1/k_2)}{k_1 - k_2}\right] - \exp\left[-k_2 \ln\frac{k_1/k_2)}{k_1 - k_2}\right] \right\}$$

$$= \alpha \left[\left(\frac{k_1}{k_2}\right)^{-k_1(k_1-k_2)} \left[1 - \left(\frac{k_1}{k_2}\right)^{-k_2+k_1)/(k_1-k_2)} \right] \right]$$

$$= \alpha \left[\left(\frac{k_1}{k_2}\right)^{-k_1(k_1-k_2)} \left[1 - \left(\frac{k_1}{k_2}\right)^{(-k_2+k_1)/(k_1-k_2)} \right] \right]$$

$$= \left(\frac{k_1 a_0}{k_2 - k_1}\right)\left(\frac{k_1}{k_2}\right)^{-k_1/(k_1 - k_2)} \left(1 - \frac{k_1}{k_2}\right)$$

$$= \frac{k_1 a_0}{k_2}\left(\frac{k_1}{k_2}\right)^{-k_1/(k_1 - k_2)}$$

$$= a_0\left(\frac{k_1}{k_2}\right)^{-k_1/(k_1 - k_2)}.$$

Case II. $k_1 = k_2 = k$

The general solution for b is

$$b = \alpha e^{-kt} + \beta t e^{-kt}$$

When $t = 0$, $b = 0$; $0 = \alpha + \beta \, . \, 0$, $\alpha = 0$. Then

$$b = \beta t e^{-kt'} = \beta\,(1 - kt) - kt$$

As in case I

$$\dot{a} = -ka$$

$$-\dot{b} - \dot{c} = -k\,(a_0 - b - c)$$

$$-\dot{b}\,kb = -k\,(a_0 - b - c)$$

When $t = 0$, $b = c = 0$; $\dot{b} = \beta = ka_0$

When $\dot{b} = 0$, $1 - kt = 0$, and $t_{max} = 1/k$.

$$\ddot{b} = -k_1k_2b \;\; 0$$

as before.

$$b_{max} = \beta\tau_{\mu\alpha\xi}\varepsilon^{-\kappa\tau_{\mu\alpha\xi}}$$

$$= ka_0 \, . \, \frac{1}{k}e^{-1}$$

$$= \frac{a_0}{e}$$

PHOTOCHEMISTRY

SOLUTION 1 : At 365.5 mμ

$$\phi = \frac{\text{number of molecules decomposed}}{\text{number } \textit{of photons} \text{ abosorbed}}$$

$$= \frac{5.18 \times 10^{18}}{10.58 \times 10^{18}} = 0.490$$

The values of for the four measurements are, respectively, 0.490, 0.483, 0.576, 0.577.

SOLUTION 2 : A=abc, where A = absorbance, a = the absorptivity of the material, b = thickness of the sample, and c = concentration.

T =I/I0 A = -\- log *T*

Solution No. 3 : At 4000mm, $A = \log \frac{1}{0.1} = 1.00; a_A = \frac{1.00}{0.001}$

$$= 1.00 \times 10^3$$

At 500mm, $A = \log \frac{1}{0.60} = 0.22; a_A = \frac{0.22}{0.001} = 0.22 \times 10^3$

Solution No. 2 : At 400mm, $A = \log \frac{1}{0.80} = 0.097; a_B = \frac{0.097}{0.005}$

$$= 19.4$$

At 500mm, $A = \log \frac{1}{0.20} = 0.097; a_B = \frac{0.699}{0.005} = 140$

Solution No. 3 : At 400nm, $1000C_A + 19.4\ C_B = \log \frac{1}{0.40} = 0.398$

At 500nm, $220C_A + 140\ C_B = \log \frac{1}{0.50} = 0.301$

Solving the last two equations simultaneously.

$$C_B = 0.00156 \text{ moles liter}^{-1}$$

$$C_{in} = 0.000368 \text{ moles liter}^{-1}$$

SOLUTION 3 : $A_{HIn} = 0.142; a_{HIn} = \frac{0.142}{6.36 \times 10^{-4}} = 223$

$A_{In^-} = 0.943; a_{In^-} = \frac{0.943}{6.36 \times 10^{-4}} = 1483$

let [H*In*] = the concentration of H*In* in the solution with I = 0.02

Let [*In*⁻] = the concentration of In⁻ in the same solution

Then

$[HIn] = 6.36 \times 10^{-4} - [In^-]$

$223[HIn] + 1483[In^-] = 0.470$

$223\,(6.36 \times 10^{-4} - \{In^-]) + 1483[In^-] = 0.470$

$0.142 - 223[In^-] + 1483[In^-] = 0.470$

$1260[In^-] = 0.328$

$[In^-] = 2.60 \times 10^{-4}$

$[HIn] = 6.36 \times 10^{-4} - 2.60 \times 10^{-4} = 3.76 \times 10^{-4}$

$pH = 8.207; [H^+] = 6.21 \times 10^{-9}$

$$K = \frac{[H+][In^-]}{[HIn]} = \frac{6.21 \times 10^{-9} \times 2.60 \times 10^{-4}}{3.76 \times 10^{-4}}$$

$= 4.29 \times 10^{-9}$

$pK = 8.368$ at I = 0.02

I	0.10	0.08	0.06	0.04	0.02
p K	8.36	8.36	8.36	8.36	8.37

SOLUTION 4 : $HIn = H^+ + In^-$

$$Ka = \frac{[HIn]}{[H^+][In^-]} = \frac{[HIn]}{[In^-]} \times \frac{1}{[H+]}$$

$$pKa = \log \frac{[HIn]}{[In^-]} + pH$$

$$A = \log \frac{1}{T}$$

where A is absorbance

Relative Concentration	pH = 4.39 *A*	Strongly basis *A*
0.8	0.699	1.268
0.6	0.538	0.959
0.4	0.355	0.644
0.2	0.175	0.333

Assuming the B'eer's law holds, that is, assuming that the concentration of the base form [In⁻] is proportional to the absorbance A, we can write, referring to the 0.8 solution, buffered at $pH = 4.39$,

$$\frac{[HIn]}{[In^-]} = \frac{1.268 - 0.699}{0.699}$$

$$pK = \log 0.814 + pH$$
$$= -0.09 + 4.39 = 4.30$$

The correspoding results for the 0.6, 0.4, and 0.2 solutions are, respectively, 4.17, 4.23, 4,34. The average of the four solutions is 4.26.

SOLUTION 5 : Rearrangement of (4):

$$k_4[HO_2]^2 = k_3[H_2O_2][OH] \quad (5)$$

Substitution of (5) into (1):

Rearrangement of (3):

$$k_1[OH][CO] = k_2[H_2O_2][CO_2H]$$

Substitution of (7) into (2):

$$2\phi\, I_a = k_3[H_2O_2][CO_2H]$$

Substitution of (7) into (6):

$$\frac{d[H_2O_2]}{dt} = -\phi I_a - k_1[OH][CO]$$

Rearrangement of (8):

$$[OH] = \frac{2\phi I_a}{k_3[H_2O_2]}$$

Substitution of (10) into (9):

$$\frac{d[H_2O_2]}{dt} = -\phi I_a - \frac{2k_1\phi I_a[CO]}{k_3[H_2O_2]}$$

Rearrangement of (11):

$$\frac{d[H_2O_2]}{dt} = -\phi I_a\left(1 + \frac{2k_1 I_1[CO]}{k_3[H_2O_2]}\right)$$

Equation (12) is the form obntained by Buxton and Walker.

SOLUTION 6 :

$$\frac{dCO_2H]}{dt} = k_1[OH]\{CO] - k_2[H_2O][CO_2H] - k_2[CO_2H]^2 = 0 \quad (2)$$

$$\frac{d[OH]}{dt} = 2\phi I_a - k_1[OH][CO] + k_2[H_2O_2] = 0 \quad (3)$$

Sum of (2) and (3):

$$2\phi I_a - k_2[CO_2H]^2 \quad (4)$$

$$[CO_2H] = \sqrt{\frac{2\phi I_a}{k_2}} \quad (5)$$

Substitution of (5) into (1):

$$\frac{d[H_2O_2]}{dt} = -\phi I_a - \frac{k_2[2\phi I_a]^{1/2}[H_2O_2]}{k_5^{1/2}} \quad (5)$$

SOLUTION 7: Absorbance = abc = A, where a = absorptivity, b, = sample path length, and c = concentration.

$$T = \frac{I}{I_0} = 10^{-abc} = e^{-2.303abc}$$

$$dT = abe^{-2.303abc}dc$$

Let

$$P = \frac{1}{c}\frac{dc}{dT} = \frac{-e^{2.303abc}}{abc}$$

p represents the relative error in c divided by the error in T.

$$\frac{dp}{dc} = \frac{-2.303abe^{2.303abc}}{abc} + \frac{e^{2.303abc}}{abc^2} = 0$$

$$\frac{-2.303}{c} + \frac{1}{abc^2} = 0$$

$$abc = \frac{1}{2.303}$$

$$T = e^{-2.303abc} = e^{-1} = 0.368$$

The error *p* is a minimum when t = 36.8 percent.

SOLUTION 8: (a) $\frac{2\text{molecules}}{\text{photon}}$

$$\times \frac{2.070 \times 10^{-5}\ cm}{6.625 \times 10^{-27}\ erg\ sec/photon \times 2.998 \times {}^{`}10^{10}\ cm\ sec^{-1}}$$

$$\times 4.184 \times 10^{7}\ erg\ cal^{-1} \times \frac{127.9\ g}{6.023 \times 10^{23}\ molecules}$$

$$= 1.85 \times 10^{-3}\ g\ cal^{-1}$$

(b) Each photon destroys one HI molecule, producing an H atom that destroys a second HI molecule in agreement with the observed quantum yield of 2.

(c) the high quantum yield of the $H_2 + Cl_2$ reaction is attributed to the self-perpetuating sequence

$Cl + H_2 \rightarrow HCl + H$ (1)

$H + Cl_2 \rightarrow HCl + Cl$ (2)

The energies of activaticn ar fairly low: $E_1 = 8$ kcal, $E_2 = 2$kcal. Thus both reactions are fast and can occur repreatedly before the chain is terminated by recombination of atoms. The H*l* decomposition would show a similarly high quantum yield if the ollowing chain were possible:

$H + HI \rightarrow HCl + H$ (1)

$H + Cl_2 \rightarrow HCl + Cl$ (2)

$E_3 = 1.5$ kcal, so that reaction (3) should be fast and it is indeed assumed as one step in the mechanism. E_4 is not given but a lower limit can be obtained from bond energies:

	$\Delta E, kcal$
$HI \rightarrow H + 1$	71
$2I_2$	−36
$1 + HI \rightarrow I_2 + H$	+35

Thus $E_4 \geq 25$ kcal; the energy required to reach the activated state cananot be less than the energy required to reach the final products. We expect reaction (4) with its high activation and l atoms will recombine much more often than they will undergo (4).

SOLUTION 9

(a) $$\frac{d[O*]}{dt} = \phi I_a - k_2[O*][O_3] - k_2[O*] \approx 0$$

$$[O*] = \frac{\phi I_a}{k_2[O_3] + k_3}$$

$$\frac{d[O]}{dt} = k_3[O*] - k_4[O][O_2][M] \approx 0$$

$$[O] = \frac{d[O_2]}{dt}$$

$$= -\phi I_a - k_2[O*][O_3] + k_4[O][O_2][M]$$

$$= -\phi I_a + (k_3 - k_2[O_3])[O*]$$

$$= \phi I_a\left(-1 + \frac{k_3 - k_2[O_3]}{k_2[O_3] + k_3}\right)$$

$$= \phi I_a\left(\frac{-2k_2[O_3]}{k_2[O_3] + k_3}\right)$$

$$\varnothing^{-1} = \phi^{-1}\left(\frac{1}{2} + \frac{k_3}{2k_2[O_3]}\right)$$

(b) $(2\phi)^{-1} = 0.538, \quad \phi = 0.93$

$$\frac{k_3}{2\phi k_2} = 0.81, \quad \frac{k_3}{k_2} = 1.51 \text{ mole liter}^{-1}$$

SOLUTION 10: (a) Rate = kI_0^n

Log rate = log k + n log I_0

Plot "log rate" vs "log I_0"

Slope = n =

(b) Rate = $k'P^m_{cl2}$

Log rate = k^1 + μ log P_{cl2}

Plot "log rate" vs "log P_{Cl2}"

Slope = m = 3/2

(c) Rate = $k''P^{3/2}_{cl2}\, I^{1/2}_0$ = rate desappearance of Cl_2 = rate of appearance of $TfCl_2$.

(d) Since $h\nu$ in this case refers to incident (rather than absorbed light, the rate of reaction (1) is $k_1 I_0 P_{cl2}$, not $k_1 I_0$ (provided $I_a \ll I_0$).

$$\frac{-dP_{Cl_2}}{dt} = k_1 I_0 P_{Cl_2} + k_3 P_{Cl_2} P_{TfCl} \quad (1a)$$

$$\frac{dP_{TfCl_2}}{dt} = k_3 P_{TfCl} P_{Cl_2} \quad (1b)$$

Making the steady state approximation for Cl, we obtain

$$k_2 P_{cl} P_{tl} = 2k_1 P_{cl2} I_0 + k_3 P_{ticl} P_{cl2} \quad (2)$$

Making the steady state approximation for $TfCl$, we obtain

$$k_3 P_{TfCl} P_{Cl_2} + k_4 P^2_{TfCl} = k_2 P_{Cl} P_{Tf} \quad (3)$$

Solving (2) and (3) simultaneously we obtain

$$k_4 P^2_{Ticl} = 2k_1 P_{cl2} I_0 \quad (4)$$

or

$$P_{TfCl} = \left(\frac{2k_1 P_{Cl_2} I_0}{k_4}\right)^{1/2}$$

Substituting this value for P_{Tfcl} (5) into (la), we obtain

$$\frac{-dP_{Cl_2}}{dt} = k_1 I_0 P_{Cl_2} + k_3 P_{Cl_2} \left(\frac{2k_1 P_{Cl_2} I_0}{k_4}\right)^{1/2} \quad (6)$$

$$= k_1 I_0 P_{Cl_2} + k''' P^{3/2}_{Cl_2} I_0^{1/2}$$

If reaction (1) is very much slower than reactions (2), (3), and (4), then the terms $k_1 I_0 P_{cl2}$ is small and m ay be neglected with respect to the term $k''' P^{3/2}_{cl2} I^{1/2}$ in equation (6), so that (6) reduced to the form

$$\frac{-dP_{Cl_2}}{dt} - k^m P_{Cl_2}^{3/2} I_0^{1/2} \quad (7)$$

Alternatively starting with equation (lb) substituting equation (5) for P_{Tfcl} , we obtain

$$\frac{dP_{TfCl_2}}{dt} = k_3 P_{TfCl} P_{Cl_2} = k^m P_{Cl_2}^{3/2} I_0^{1/2}$$

(e) If their concentrations have no effect on the reaction rate.

(f) One would have to work lout the new rate law based on the reverse of reaction (3) in addition to the other reactions and compare this law with the experimental results.

(g) Free redicals are involved.